FRCPath Part 1:
Examination
Preparation Guide

First edition

FRCPath Part 1: Examination Preparation Guide

Dr S. Steele
Dr S. O'Connor

Academic Medical Press

Academic Medical Press, a division of Academic Medical Consulting (Publisher).
Nottingham, UK.

Academic Medical Consulting.
Ebury Road, Carrington, Nottingham NG5 1BB
Somniare audemus

First published 2011

ISBN 978-0-9566443-1-2

Further copies can be obtained from: http://www.amazon.com
http://www.lulu.com

Preface to the first edition

FRCPath Part 1: Examination Preparation Guide, is intended for junior pathologists preparing to attempt the first phase of the Royal College of Pathologists membership examination. It arises out of the authors' experiences of teaching histopathology to trainees prior to this examination. A common complaint from our colleagues has been the lack of a substantial publication to aid preparation for the examination. This publication is intended to remedy that deficit.

The book is a collection of mock examinations similar in style and standard to the FRCPath Part 1 examination. Students are advised to augment their ongoing revision with the questions in this book. We hope that this approach will improve both the focus and clarity of their revision and the quality of their exam performance.

The authors believe that this book will improve the candidate's chances of success in their upcoming FRCPath Part1 examination.

Nottingham, August 2010   Sinclair Steele and Simon O'Connor

How to use this book

After revising the appropriate material, most candidates will benefit from attempting the Mock examinations in this text. The questions have been designed to mimic the actual FRCPath Part 1 examination in standard, style and organization. Each paper begins with 50 single best answer (SBA) questions and is followed by 75 extended matching questions (EMQ). 3 hours are allowed for each paper. The precise pass mark for the real examination is set by the Royal College of Pathologists. However, the authors would advise that a candidate answering more than 66% of the questions correctly would be likely to achieve a pass standard.

The size of paper used and spacing of the questions is also similar to that of the FRCPath Part 1 examination; allowing the student a more realistic impression of exam conditions. The thorough candidate should consider making notes in the spaces provided and reviewing these notes after they have finished the paper, as part of their own internal audit.

Trainees are aware that part of the purpose of any postgraduate medical examination is to differentiate between the candidates. Hence questions are included in such examinations to confirm that the student not only has the most appropriate level of knowledge but also has the good judgement to use such knowledge appropriately. Accordingly, some of the included questions may have either unfamiliar phrasing or an unfamiliar context, in order to take the candidates out of their comfort zones and make appropriate critical judgements. This is in contrast to most trainees' experience of day-to-day pathology, which may tend towards the routine. The intelligent trainee pathologist should be able to discern the key information in unusual or unfamiliar questions and, nevertheless, select the best answers.

CONTENTS

Preface   4

How to use this book   5

Examination Papers:

Paper 1   7 - 33
Answers   34 - 49

Paper 2   50 - 76
Answers   77 - 92

Paper 3   93 - 118
Answers   119 - 136

Paper 4   137 - 161
Answers   162 - 177

Paper 5   178 - 202
Answers   204 - 217

Page(s)

PAPER 1

1) An autopsy of a patient who died of a suspected myocardial infarction revealed macroscopic features of coagulative necrosis and a sparse neutrophilic infiltrate on histological examination. The coroner has asked for an estimate of the period between the myocardial infarction and death. Which is the most accurate answer?

(A) 0 – 12 hours
(B) 12 – 24 hours
√(C) 1 – 3 days
(D) 3 – 5 days
(E) 4 – 7 days

2) A trainee pathologist is attempting to interpret a report on cerebrospinal fluid from a patient with suspected meningitis. The pathologist has difficulty interpreting the results because he cannot remember the normal components of cerebrospinal fluid. Which of the following is the most accurate description of the cellular components of normal CSF?

(A) Neutrophils, lymphocytes and monocytes
(B) Lymphocytes, monocytes and choroid plexus cells
(C) No cells (acellular)
√(D) Lymphocytes and monocytes only
(E) Red blood cells and lymphocytes

3) A widely used system of classification and coding of surgical biopsy pathology is the SNOMED. Which category of SNOMED most directly relates to aetiology?

(A) T
(B) M
(C) D
√(D) E
(E) F

4) A surgical biopsy taken from a patient with a renal mass showed a malignant tumour. Histologically the tumour was a recurrence of a renal tumour that had occurred 5 years previously. At the time it had been described as "malignant with low metastatic potential." Which of the following renal tumours is most likely to have been the primary?

(A) Clear cell carcinoma
(B) Collecting duct carcinoma
(C) Oncocytoma
√(D) Papillary renal cell adenocarcinoma
(E) Angiomyolipoma

5) "A Chief Executive-led management structure responsible for ensuring quality of patient care." Refers to:

√(A) Clinical governance

(B) Audit
(C) EQA
(D) Revalidation
(E) Accreditation

6) A glomus tumour of the skin originates from which of the following cells/tissues?

(A) Vascular endothelial cells
✓ (B) Smooth muscle cells
(C) Epidermal squamous cells
(D) Sebaceous glands
(E) Eccrine cells

7) A patient presents with a pigmented skin lesion of long standing. The lesion is 6 mm in diameter. An excisional biopsy of the lesion revealed junctional and dermal melanocytic cell proliferation. The rete ridges were extended with some crosslinking by horizontal spindle shaped melanocytes. No pagetoid spread or uniform cellular atypia was noted. Lamellar fibroplasia was noted adjacent to the rete ridges. A perivascular lymphocytic response was noted in the superficial dermis. Adjacent solar elastosis was present. The lesion is most likely to be:

(A) Malignant melanoma
(B) Lentigo simplex
(C) Compound naevus
✓ (D) Dysplastic naevus
(E) Spitz naevus

8) A prostate biopsy shows malignant glandular cells organised predominantly as packed angular glands with a lesser component of cribriform architecture. Which of the following is the most accurate Gleason grade?

(A) 2 + 2
(B) 2 + 3
(C) 5 + 5
✓ (D) 3 + 4
(E) 4 + 3

9) Which of the following clinical scenarios require referral of the autopsy to the local coroner?

(A) A man with a history of angina saw his doctor 10 days ago and collapsed and died whilst running for a bus.
✓ (B) An elderly woman died in a nursing home after fracturing her right neck of femur.
(C) A longstanding smoker with a history of lung cancer and chemotherapy dies at home.
(D) A patient with a history of TB dies with weight loss and respiratory failure.
(E) A young man with a history of HIV infection has an episode of sepsis and dies.

10) A nephrectomy specimen contains a necrotic 55mm tumour confined to the kidney. There is no extracapsular spread. Vascular invasion is present that does not extend to the resection margin. What is the T stage of this tumour?

(A) T1
(B) T2
(C) T3a
√(D) T3b
(E) T3c

11) A right cheek mass in a 40 year old male patient is biopsied. Histology reveals the following components of the tumour; chondromyxoid stroma, ducts and bland epithelial cells. Which of the following is most likely to be the tumour type?

(A) Warthin's tumour
(B) Polymorphous low grade adenocarcinoma
√(C) Pleomorphic adenoma
(D) Adenoid cystic carcinoma
(E) Squamous cell carcinoma

12) A lymph node biopsy taken from the posterior triangle of the neck in a 35 year old man showed an encapsulated mass with a collection of small lymphoid cells. Focal accumulations of histiocytes amounting to non-caseating granulomata are present. Aggregates of monocytoid cells with pale cytoplasm are also noted. What is the likely aetiological agent?

√(A) Toxoplasma
(B) Mucor
(C) EBV
(D) HIV
(E) TB

13) A patient with suspected Sjogren's syndrome has a biopsy taken to confirm the clinical diagnosis. Which is a likely site of the biopsy?

(A) Submandibular gland
(B) Sublingual gland
(C) Maxillary gland
√(D) Parotid gland
(E) Posterior triangle

14) A 65 year old with a history of smoking and working in the building industry goes to see his GP because of a persistent cough. He also complains of chest pain. What is his most likely underlying pathology?

√(A) Asbestosis
(B) Lung adenocarcinoma
(C) Malignant mesothelioma
(D) Emphysema
(E) Pneumonia

15) A testicular tumour biopsy that was macroscopically a homogenous cream coloured lesion and microscopically seminomatous also had a secondary component. This secondary component contained pleomorphic cells with a pseudoglandular architecture that was CD30 positive. The second component was most likely to be:

(A) Choriocarcinoma
✓ (B) Embryonal carcinoma
(C) Yolk sac tumour
(D) Leydig tumour
(E) Intratubular germ cell neoplasia

16) A 28 year old woman was found to have multiple polyps on colonoscopy. Her father died of colon cancer at age 50. She is most likely to have mutations in the:

✓ (A) APC gene
(B) p53 gene
(C) Rb gene
(D) HNPCC gene
(E) C-myc

17) A breast biopsy shows pleomorphic ductal cells that do not extend beyond the duct wall and are associated with coarse calcification. Mitotic activity is present. What is the most likely tumour type?

✓ (A) Ductal carcinoma in situ
(B) Lobular carcinoma
(C) Mucinous carcinoma
(D) Medullary carcinoma
(E) Papillary carcinoma

18) What was the original name for the HER2 gene?

(A) c-myc
(B) BRCA1
(C) BRCA2
(D) Herceptin
✓ (E) c-erb-B2

19) Which organization in the UK deals with the collection of CPD points for revalidation?

(A) DOH
(B) GMC
(C) BMA
✓ (D) RCPath
(E) CPA

20) An FNA specimen extracted from a breast lump in a 35 year old woman was examined as a cytological smear. The polygonal cells present were reported as

showing suspicious atypia. Which of the following classifications is most appropriate for this description?

(A) C1
(B) C2
(C) C3
√(D) C4
(E) C5

21) A 45 year old male suffers a brain injury with tonsillar herniation of the brainstem. At post mortem the following features will be present except:

(A) Cerebral oedema
(B) Indentation around the tonsillar region
√ (C) Diffuse axonal injury
(D) Papilloedema
(E) Acute axonal injury

22) A 30 year old woman gave birth to a normal male child after a difficult 36 hour labour. 12 hours after the delivery the woman complained of shortness of breath, suffered a hypotensive episode, seizure, coma and death. What is the most likely underlying pathology?

(A) Pulmonary embolism
(B) Undiagnosed phaechromocytoma
(C) Cervical tear and postpartum haemorrhage
(D) Pre-eclampsia
√(E) Amniotic fluid embolism

23) A 4 month old baby presented with vomiting and a stiff neck. A diagnosis of meningitis was made. The most likely causative agent was:

(A) Neisseria meningitides
(B) Streptococcal pneumonia
√ (C) Haemophilus influenza
(D) Staphylococcus aureus
(E) Streptococcus faecalis

24) A bladder biopsy showing a mixed inflammatory infiltrate contained areas of calcification. The foci were intracellular and extracellular lamellated structures containing concentric calcification, which were centrally dark staining. Identify the most likely structures:

(A) Psammoma bodies
√(B) Michaelis – Guttmann bodies
(C) Metastatic calcification
(D) Dystrophic calcification
(E) Schistosomiasis

25) A 30 year old man with a history of immunosuppression suffers from a brain infection. The most likely cause of his infection is:

(A) Coccidiomycosis
(B) CMV
(C) Staphylococcus aureus
(D) Streptococcus
(E) EBV

26) A 28 year old woman presents with a hard lump in her thyroid gland which was diagnosed as a medullary carcinoma of the thyroid. On further questioning, the patient reveals that her father had an operation to remove an adrenal gland a year previously. Her brother had a pituitary adenoma. Identify the genetic condition that this woman is most likely to be suffering from:

(A) Gardner's syndrome
(B) Marfan's syndrome
(C) Multiple endocrine neoplasia
(D) Tuberous sclerosis
(E) Von Recklinghausen's disease

27) A two month old male child sleeps at home in the same bed as his mother. The child has been well since birth and has no notable medical history. The developmental milestones are normal. The mother wakes up one morning to find the infant unresponsive. The GP is called and the infant is pronounced dead. There are no signs of maltreatment of the infant. The skeletal survey is negative. What is the most likely diagnosis?

(A) Pneumonia
(B) Croup
(C) Meningitis
(D) SIDS
(E) Epilepsy

28) You are asked to carry out a post mortem examination on a 50 year old female who has been on long term renal dialysis. The mortuary technician asks for advice on the most hazardous part of the post mortem examination of this patient. Which of the following procedures carries the greatest risk?

(A) Opening the thoracic cavity.
(B) Opening a lung.
(C) Opening the abdominal cavity.
(D) Opening the intestines.
(E) Opening the skull.

29) A 6 month old infant with a history of vomiting and bloody diarrhoea has died and a post mortem was performed. During the autopsy part of the large intestine was found to be ischaemic. The most likely diagnosis is:

(A) Hirschsprung's disease

13

√(B) Necrotising enterocolitis
(C) Wilms tumour
(D) Crohn's disease
(E) Diverticulitis

30) A young woman is killed in a car crash and a coroner's autopsy is performed. Signs of a critical head injury are found. The pathologist carrying out the post mortem has noted hilar lymphadenopathy and would like to take a sample for histology. Consent is required from:

(A) Next of kin
√(B) Coroner
(C) Patient's doctor
(D) Coroner's officer
(E) Accident and emergency consultant

31) A young woman suffering from dysmenorrhea is examined under ultrasound and a bulky uterus is noted. A hysterectomy is performed and the macroscopic examination of the uterus reveals several large cream coloured well demarcated nodules that are located in the myometrium. What is the most likely underlying pathology?

(A) Leiomyosarcoma
(B) Endometrial stromal tumour
(C) Endometriosis
√(D) Leiomyoma
(E) Thecoma

32) The cervical screening program requires storage of cervical smears. According to the requirements of the screening program, how long should the slides be kept?

(A) 3 years
√(B) 10 years
(C) Until the woman is 65.
(D) Until 3 negative smears are performed.
(E) Until colposcopy is performed.

33) A 40 year old daughter, presents with a colonic polyp that on investigation is shown to be an invasive adenocarcinoma. Her father died from a large intestinal adenocarcinoma at age 50. Which familial syndrome is she most likely to have inherited?

√(A) HNPCC
(B) FAP
(C) Marfan's
(D) Muir-Torre Syndrome
(E) Gardner's syndrome

14

34) A young woman with a cervical mass had it excised at colposcopy. Histological examination demonstrated CIN I. What advice should her physician give to decrease the risk of a further cervical neoplasm?

(A) Refrain from sexual activity.
✓(B) Use a barrier method of contraception.
(C) Stop smoking.
(D) Stop drinking.
(E) Repeated colposcopy.

35) A post mortem was performed on a 35 year old man who died suddenly from a presumed cardiac lesion. Macroscopically the heart was enlarged. Histological revealed fibroadipose nodules and inflammation in the myocardium. What is the most likely underlying pathology?

(A) Alcoholic cardiomyopathy
(B) Dilated cardiomyopathy
(C) Hypertrophic obstructive cardiomyopathy
✓(D) Arrhythmogenic cardiomyopathy
(E) Restrictive cardiomyopathy

36) A Cantonese woman who does not speak English attends an accident and emergency clinic and requires a surgical procedure. In order to perform the procedure:

✓(A) Send the woman home and give her a date to return with an interpreter present.
(B) Use sign language to indicate procedure.
(C) Get a nurse to hold her steady and do FNA (trying not to get a needle-stick!)
(D) Phone surgeon and arrange urgent open biopsy.
(E) Tru-cut biopsy the lesion.

37) What is the most accurate definition of emphysema?

✓(A) Abnormal and permanent enlargement of the airspaces distal to the terminal bronchiole, accompanied by destruction of the walls and without obvious fibrosis.
(B) Abnormal and permanent enlargement of the airspaces distal to the respiratory bronchiole, accompanied by destruction of the walls and without obvious fibrosis.
(C) It is present in any patient who has persistent cough with sputum production for at least 3 months in at least 2 consecutive years.
(D) Chronic relapsing inflammatory disorder characterized by hyperreactive airways leading to episodic, reversible bronchoconstriction, owing to increased responsiveness of the tracheobronchial tree to various stimuli.
(E) Abnormal and permanent enlargement of the acinus airspaces.

38) Which of the following best describes the disease aetiology of a child born with spina bifida?

(A) Folate deficiency
(B) Chromosomal abnormality
✓(C) Multifactorial
(D) Environmental factors

(E) Single gene defect

39) A patient has a suspected crystal arthropathy of a major joint. A sample of synovial fluid is taken and sent to the laboratory services. It is ultracentrifuged. What is the best next step with respect to the handling and histological examination of the specimen?

(A) Wash the specimen in saline.
√(B) Examine under light microscopy with plane polarized light.
(C) Dark field microscopy.
(D) Wash the specimen with alcohol.
(E) Decalcify the specimen.

40) A renal tumour is excised from a 45 year old man. Histological examination of this 3cm mass shows muscle, vessels and fat. Immunohistochemistry revealed HMB45 positivity. What is the likely diagnosis?

(A) Malignant melanoma
(B) Renal cell carcinoma
(C) Angiolipoma
(D) Myolipoma
√(E) Angiomyolipoma

41) A male patient presents with a 900g adrenal tumour. He has signs and symptoms consistent with Cushing's syndrome. The tumour is resected and examined under light microscopy. Which of the following would be the poorest prognostic sign?

(A) Clear cell morphology.
(B) 1 mitosis in 10hpf.
(C) Diploid DNA content.
√ (D) Large size.
(E) ACTH production.

42) A processed slide is delivered to a consultant with an associated request form that indicates a colonic sample. Under the microscope the slide demonstrates endometrial tissue. What is the first thing that the consultant might do to correct this error?

√(A) Check handwritten annotation under label.
(B) Check block.
(C) Phone surgeon for history of endometriosis.
(D) Re-cut slide.
(E) Throw the slide away and get a re-cut.

43) A 50 year old woman presents with an ovarian mass. The resected mass is 13cm in diameter and has solid and cystic areas. The solid areas are yellow-gray in colour. Immunohistochemistry was positive for:

(A) Alphafetoprotein
√ (B) Inhibin
(C) CEA

(D) CA 125
(E) OM-1

44) A 40 year old woman presents with an 8cm small bowel tumour. On resection and histological examination the tumour was positive to histological challenge with CD117, CD34 and SMA. What is the most likely diagnosis?

✓(A) GIST
(B) Leiomyoma
(C) Leiomyosarcoma
(D) Lymphoma
(E) Carcinoid

45) A 60 year old male alcoholic is found dead at home. At post mortem his lungs were heavy, amorphous and congested. On histological examination his lungs contained unidentifiable material, which elicited an inflammatory response. What is the likely cause of death?

(A) Bronchopneumonia
(B) Lobar pneumonia
✓(C) Aspiration pneumonia
(D) Stroke
(E) Arrhythmia

46) Which of the following is not associated with HNPCC colon cancer?

(A) hMLH1 mutation,
(B) hMSH2 mutation,
✓(C) v-sis mutation,
(D) microsatellite instability
(E) hMSH6

47) A four year old child presents with signs and symptoms of renal failure. On biopsy the histology shows normal glomeruli, tubules, vessels and interstitium. Immunofluorescence revealed no immunoglobulin or complement deposits. Electron microscopy demonstrated uniform and diffuse effacement of foot processes.
What is the likely diagnosis?

(A) Membranoproliferative glomerulonephritis
(B) Focal segmental glomerulosclerosis
(C) Lupus nephritis
(D) IgA disease
✓(E) Minimal change disease

48) A 25 year old woman had a lesion in her right leg that was biopsied. The histopathologist described the lesion as benign – unfortunately the woman presented with lung metastases 4 years later. What was the original lesion most likely to have been?

(A) Giant cell tumour of tendon sheath

√(B) Giant cell tumour of bone

(C) Leiomyosarcoma

(D) Rhabdomyosarcoma

(E) Leiomyoma

49) A 50 year old smoker has a pleural effusion that is tapped and sent for cytopathological assessment. The cells are polygonal with mild pleomorphism and exist as small clumps and free cells, some with signet ring morphology. Immunocytochemistry is performed and the cells are positive to AE1/AE3, BerEp4 and negative for calretinin. What is the diagnosis?

(A) Mesothelioma

√(B) Adenocarcinoma

(C) Squamous cell carcinoma infiltrating the pleura

(D) Lymphoma

(E) Small cell carcinoma

50) A 29 year old male intravenous drug user presents with the signs and symptoms of hepatic impairment. This smoker undergoes a liver biopsy and the histology reveals cirrhotic features. What is the most likely pathogenesis of his condition?

(A) HIV infection

(B) TB infection

√(C) Hepatitis C infection

(D) Alcoholic hepatitis and cirrhosis.

(E) EBV infection

Using the clinical scenarios below choose the most likely diagnosis from the list of conditions given:

A) Sinus histiocytosis
B) Chronic lymphocytic leukaemia
C) Mantle cell lymphoma
D) Burkitt's lymphoma
E) Diffuse Large B cell lymphoma

F) Mycosis fungoides
G) Multiple myeloma
H) Follicular lymphoma
I) Follicular hyperplasia
J) Marginal zone lymphoma

K) Hodgkin's lymphoma
L) Castleman's disease
M) Acute lymphoblastic leukaemia
N) Multiple myeloma

51) A 20 year old woman presents with a malignancy with monotonous large lymphoid cells with a high mitotic rate. Immunohistochemistry shows:
CD3 neg. CD5 neg. CD10 pos. CD20 pos. CD79a pos. CD23 neg. CD43 neg.
c-myc pos.   D

52) A 60 year old man presents with a malignancy composed of uniform small to medium sized lymphoid cells. Immunohistochemical staining is:
CD 3 neg. CD5 pos. CD10 neg. CD23 pos. CD 43 pos. CD79a pos. cyclin D1 neg.
     B

53) A 55 year old woman presents with a malignancy composed of medium sized centrocytoid lymphoid cells and larger nucleated cells. A mixed diffuse and nodular architecture is noted. Immunohistochemical staining is:
CD3 neg. CD5 neg. CD10 pos. CD20 pos. CD79a. pos. CD43 neg. CD23 neg.
     H

54) A 60 year old man presents with a malignancy composed of medium sized lymphoid cells with dispersed chromatin. Immunohistochemical staining is:
CD3 neg. CD5 pos. CD20 pos. CD23 neg. CD79a pos. Cyclin D1 pos.
     C

55) A 62 year old woman presents with a malignancy composed of large lymphoid cells with prominent nucleoli.
CD3 neg. CD5 pos. CD20 pos. CD79a pos. CD10 pos. CD43 neg. bcl-2 pos.
     E

19

A) Bullous pemphigoid   F) Erythema multiforme   J) Bullous amyloidosis
B) Dermatitis herpetiformis   G) Bullous amyloidosis   K) Herpes simplex
C) Pemphigus vulgaris   H) Hailey-Hailey disease   L) Epidermolysis bullosa
D) Pemphigus foliaceus   I) Porphyria cutanea   acquisita
E) Dermatitis artefacta   tarda

For each of the clinical scenarios below choose the best answer from the above list:

56) A 70 year old woman with numerous subepidermal blisters containing eosinophils.

A

57) A 17 year old woman presented with an irregular distribution of bizarrely shaped intraepidermal blisters in her hands and heels. They all appeared within the last 24 hours. The clinical history revealed no prior signs or symptoms.

E

58) A 40 year old man has a subepidermal blister containing neutrophils and eosinophils that reveals IgA under direct immunofluorescence.

B

59) This intraepidermal blister has a "crumbling brick wall" appearance histologically.

H

60) A 50 year old woman presents with a crop of acellular subepidermal blisters. Direct immunofluorescence reveals IgG and C3 deposition.

L

Using the clinical scenarios below choose the most likely diagnosis from the list of conditions given:

A) Parkinson's disease F) Stroke   K) Meningioma
B) Huntingdon's chorea G) Multiple Sclerosis
C) Alzheimer's disease H) Glioblastoma multiforme
D) Multi-infarct dementia I) Meningitis
E) Creutzfeldt-Jacob disease J) Encephalitis

61) A 75 year old man had longstanding dementia. His brain showed:
Pigmented substantia nigra, normal cerebellum, normal pons, normal hippocampus and basal ganglia, severe diffuse cortical atrophy.

C

62) This 65 year old man with progressive dementia and a festinant gait had the following findings:
Mild cortical atrophy, normal cerebellum, normal pons, normal hippocampus and basal ganglia, pale substantia nigra.

A

63) A 70 year old man had a stepwise deterioration in cognitive function over several years. The brain showed minor cortical atrophy, multiple small slit-like spaces less than 1mm in length in the cerebellum, hippocampus, basal ganglia and pons. The substantia nigra is pigmented.

D

64) A 35 year old man presents with a history of rapidly progressing dementia.
His brain shows a normal cortex, spongiform change in the cerebellum, normal hippocampus, basal ganglia and pons. The substantia nigra is pigmented.

E

65) A 45 year old man with dementia and a movement disorder, had blocks that showed atrophy of the caudate nucleus and putamen and globus pallidus. Mild cortical atrophy, normal cerebellum, pigmented substantia nigra, normal basal ganglia, hippocampus and pons.

B

Match the histological/cytological description to the most likely diagnosis chosen from the list of conditions given below:

A) Nodular colloid goitre   E) Follicular variant of papillary carcinoma
B) Thyroid cyst   F) Hashimoto's thyroiditis   I) Follicular adenoma
C) Papillary carcinoma   G) B-cell lymphoma   J) Follicular carcinoma
D) Hurthle cell neoplasm   H) Anaplastic carcinoma

66) A 50 year old woman presents with an enlarged thyroid. A fine needle aspirate reveals large polygonal cells with blue-grey cytoplasm and indistinct cytoplasmic boundaries with background lymphocytes. The lymphocytes are CD20 and CD79a positive.

G

67) A 40 year old woman presents with a thyroid nodule in the left lobe. A biopsy reveals a locus with follicular architecture and cells showing optically clear nuclei, nuclear grooves and psammoma bodies.

E

68) A 65 year old woman presents with a very large thyroid gland. Fine needle aspiration cytology reveals abundant thick and thin colloid, haemosiderin laden macrophages and follicular microbiopsies.

A

69) A 65 year old man presents with a rapidly enlarging neck mass. A fine needle aspirate demonstrated a cellular smear with cytologically bizarrely shaped cells and a dirty background.

H

70) A 35 year old woman presents with a neck mass. Fine needle aspiration cytology revealed sparse colloid, abundant macrophages, few follicular cells and blood.

B

A) Stroke

B) Ischaemic heart disease

C) Arrhythmogenic
   cardiomyopathy

D) Aortic stenosis

E) Mitral prolapse

F) Dilated cardiomyopathy

G) Wolff-Parkinson-White
   syndrome

H) Aortic rupture

I) Hypertrophic obstructive
   cardiomyopathy

J) LQT syndrome

K) Myocarditis

L) Commotio cordis

M) Senile myocardium

Choose the best answer from the list above that applies to the scenarios described below:

71) A post mortem heart shows fibroadipose infiltration of the right ventricle.

C

72) A post mortem heart was enlarged with an increased left ventricular weight. Histology showed myocyte disarray.

B

73) A patient with a chronic history of chest pain collapses and dies. At post mortem the bore of the left anterior descending artery was diminished to 10%.

I

74) A 30 year old male driver dies in a road traffic accident. At post mortem, bruising over his sternum is noted but his heart and major vessels appear normal.

L

75) A retired woman in a nursing home dies in her sleep. She was not seen by her doctor for 6 months before her death. Autopsy reveals a soft small brown heart.

M

Match the following genes/gene products/genetic changes to the given disease states:

A) t(14:18)   E) KVLQT1   I) hMutSα
B) hMLH1   F) PiZZ   J) Rb
C) t(11;22)   G) BRCA1   K) WT-1
D) APC   H) alk1   L) NF-1

76) Follicular lymphoma

A

77) Alpha-1 antitrypsin deficiency

F

78) HNPCC

B

79) Osteosarcoma

J

80) Ewing's Sarcoma

C

24

A) CIN1　　　　　　E) Tubal metaplasia　　　I) Invasive adenocarcinoma
B) CIN3　　　　　　F) H-CGIN　　　　　　J) Radiation atypia
C) Endometriosis　　G) L-CGIN　　　　　　K) Endocervicosis
D) Squamous cell　　H) HPV infection　　　　L) Condyloma acuminatum
　　carcinoma

Relate the histological/cytological descriptions below to the most fitting pathologies above.

81. The smear contains clusters of cells showing low grade dyskaryosis.

A

82. The smear shows extensive background neutrophils and debris with sparse elongate atypical orangeophyllic squamous cells.

D

83. Cone biopsy reveals exophytic squamous lesions that show clear para-nuclear halos, binucleate cells and cellular keratinisation. Focal low grade dyskaryosis is present.

L

84. Cone biopsy reveals non-invasive glands in the endocervix that show hyperchromasia, nuclear stratification, abnormal mitoses and an intestinal-type architecture.

F

85. A peritoneal biopsy contains glands with columnar epithelium, oval nuclei with sharp nuclear membranes and no nucleoli. An associated stroma and granular brown pigment is noted.

C

A) Respiratory Syncytial Virus E) Neisseria Meningitides I) E. coli
B) Toxoplasma gondii F) Haemophilus influenzae J) Varicella zoster
C) Cytomegalovirus G) Streptococcus pneumoniae K) Beta haemolytic
D) TB H) Rubella virus streptococcus

Relate the clinical scenarios below to the most likely aetiological agent above:

86) A feverish 5 year old child is dyspnoeic with inspiratory stridor. The child is part of a poor inner city family.

87) Two weeks after a sore throat a 7 year old girl complains of joint pain.

88) A newborn infant is feverish, jaundiced, drowsy and photophobic.

89) A 4 year old child is anaemic and develops acute renal failure.

90) A neonate is born with hydrocephalus and vision problems.

A) Adenocarcinoma
B) Body of hyoid
C) Soft palate
D) Plasmacytoma
E) Epiglottis

F) Nasal polyp
G) Thyroid cartilage
H) Squamous cell carcinoma
I) Middle turbinate
J) Olfactory neuroblastoma

K) Intraepithelial tumour
L) Level one nodes
M) Level two nodes
N) Level three nodes
O) Thyrohyoid membrane

For each of the descriptions below select the most appropriate answer A) – O) from the list above.

91) The structure that marks the superior boundary of the larynx.

E

92) Structure in the nasal cavity that contains mature bone.

I

93) Melanoma metastasis causing submental lymphadenopathy.

L

94) Focal mucosal protrusion lined by pseudostratified epithelium with an oedematous stroma containing eosinophils.

F

95) Dysplastic laryngeal epithelium with a disrupted basement membrane and descending atypical epithelial cells with orangeophilic cytoplasm.

H

A) Type I collagen   F) Diaphysis   K) Sequestrum
B) Type II collagen   G) Epiphysis   L) Involucrum
C) Type III collagen   H) Epiphyseal cartilage M) Osteogenesis imperfecta
D) Type IV collagen   I) Cortical bone   N) Paget's disease of bone
E) Metaphysis   J) Cancellous bone   O) Marfan's disease

Choose the correct answer from A) – O), for each of the descriptions below:

96) Type of collagen present in bone.

97) Site of the growth plate.

98) The marrow of the diaphysis of the femur in a normal adult.

99) Non-viable bone.

100) Prone to chalk-stick fractures.

28

A) Neuroblastoma
B) Lymphoblastic leukaemia
C) Hirschprung's disease
D) Synovial sarcoma
E) Rhabdomyosarcoma
F) Wilms tumour
G) Ewing's sarcoma
H) Renal sarcoma
I) Primitive neuroectodermal tumour
J) Desmoplastic small round cell tumour
K) Osteosarcoma

Choose the pathological entities above which best match the descriptions below:

101) This apparently normal rectal biopsy showed an increased acetylcholinesterase content.

C

102) Biopsy of an abdominal mass showed primitive hyperchromatic cells with a high nuclear to cytoplasmic ratio, a stroma with bland spindle cells and a mitotically active pleomorphic acinar component.

F

103) The tumour contained numerous Homer-Wright rosettes.

A

104) A 7 year old boy's femur shows radiological evidence of bone destruction adjacent to the metaphyseal plate with an evident sunburst pattern.

K

105) This monophasic spindle cell lesion has a herringbone architecture and is CD99 and focally CAM5.2 positive.

D

A) Barrett's oesophagus   E) Kimmelstiel-Wilson disease
B) FAP   F) Coeliac disease   I) Carcinoid syndrome
C) HNPCC   G) Stiff-man syndrome   J) MEN
D) Cushing's syndrome   H) Ovarian carcinoma   K) Zollinger-Ellison syn.

Which of the diseases above most applies to the descriptions below?

106) Small cell carcinoma of the lung.

D

107) An individual with coughing, wheezing, flushing and diarrhoea.

I

108) Adenocarcinoma in the lower oesophagus.

A

109) T cell lymphoma of the small intestine.

F

110) Recurrent gastric ulceration.   *K*

A) Carbon monoxide Hb%   E) 15 lung blocks   I) HIV serology
B) Hair for phenytoin   F) Blood MDMA   J) HBV serology
C) 4 lung blocks   G) Blood mast cell tryptase
D) 10 lung blocks   H) Blood fentanyl

For each of the scenarios below, indicate the investigations that would be most useful:

111) At 2 am in a nightclub a previously healthy 19 year old boy complains of feeling warm, collapses and dies.

F

112) While eating a meal in a restaurant, a young woman with no history of heart or lung disease complains of being acutely short of breath and starts to wheeze. She collapses and dies within 15 minutes. Post mortem histology shows evidence of a myocardial infarction.

G

113) An impulsive 22 year old man, with a history of epilepsy, collapses at work, fits and dies.

B

114) An autopsy on a 75 year old ex-miner reveals ischaemic heart disease, pleural plaques and a fibrotic lung.

E

115) A 42 year old man is found dead in his car, in his own garage. The engine is still running.

A

A) Asthma
B) Emphysema
C) Bronchitis
D) Bronchiolitis

E) Bronchopneumonia
F) Lobar pneumonia
G) Adenocarcinoma
H) Coal workers' pneumoconiosis

I) Wegener's granulomatosis
J) Pulmonary tuberculosis
K) Mesothelioma
L) Adult respiratory distress syndrome

Which of the diseases above most closely matches the pathological descriptions below?

116) The pleural tissue is infiltrated by an epithelioid tumour that is calretinin and thrombomodulin positive and DPAS negative.

K

117) The lung blocks reveal extensive interstitial fibrosis, without a significant inflammatory infiltrate. Nodules with a periphery of macrophages containing anthracotic dust and central haphazardly arranged collagen, are present. The largest nodule is 3cm in maximum extent.

H

118) Lung histological examination reveals hyaline membranes, type II epithelial cell proliferation, widespread oedema and congestion.

L

119) Microscopic examination shows hypertrophy of bronchial wall muscle, submucosal gland hypertrophy and an inflammatory infiltrate with prominent eosinophils.

A

120) A block from the upper lobe of the lung shows caseating granulomata.

J

A) Classical seminoma E) Leydig cell tumour I) Anaplastic carcinoma
B) Teratoma differentiated F) Lymphoma   J) Squamous cell carcinoma
C) Embryonal carcinoma G) Small cell carcinoma K) Choriocarcinoma
D) Yolk sac tumour   H) Malignant melanoma

Which of the diseases above most closely matches the pathological descriptions below?

121) The testis is infiltrated by an epithelioid malignancy that has prominent eosinophilic nucleoli and associated macrophage bound brown pigment.

H

122) The testis contains sheets of uniform large oval/round cells with clear cytoplasm and distinct cell membranes. Nucleoli are present. The tumour is PLAP positive.

A

123) A lymph node from the left groin shows a subcapsular infiltration by epithelioid cells arranged in sheets and small groups with small round nuclei and abundant eosinophilic cytoplasm. Moderate hyperchromasia, nuclear pleomorphism and mitotic activity are noted. The tumour is CK14 positive and PLAP negative.

J

124) The testis is infiltrated by a tumour with prominent Schiller-Duval bodies, which also demonstrates eosinophilic hyaline globules and is AFP positive.

D

125) The testis displays a well circumscribed mass containing keratinizing squamous epithelium, bone and large intestinal tissue.

B

ANSWERS

1) C	43) B	85) C
2) D	44) A	86) F
3) E	45) C	87) K
4) D	46) C	88) D
5) A	47) E	89) I
6) B	48) B	90) B
7) D	49) B	91) E
8) D	50) C	92) I
9) B	51) D	93) L
10) D	52) B	94) F
11) C	53) H	95) H
12) A	54) C	96) A
13) D	55) E	97) H
14) C	56) A	98) J
15) B	57) E	99) K
16) A	58) B	100) N
17) A	59) H	101) C
18) E	60) L	102) F
19) D	61) C	103) A
20) D	62) A	104) K
21) C	63) D	105) D
22) E	64) E	106) D
23) C	65) B	107) I
24) B	66) G	108) A
25) B	67) E	109) F
26) C	68) A	110) K
27) D	69) H	111) F
28) A	70) B	112) G
29) B	71) C	113) B
30) A	72) I	114) E
31) D	73) B	115) A
32) B	74) L	116) K
33) A	75) M	117) H
34) B	76) A	118) L
35) D	77) F	119) A
36) A	78) B	120) J
37) A	79) J	121) H
38) C	80) C	122) A
39) B	81) A	123) J
40) E	82) D	124) D
41) D	83) L	125) B
42) A	84) F	

Paper 1 - Answers

1) **C**
The features of macroscopic coagulative necrosis will begin to appear at 18-24 hours, with the palor becoming marked at **24-72 hours**. The neutrophilic infiltrate is evident at 4 – 72 hours.

2) **D**
Normal cerebrospinal fluid contains only **lymphocytes and monocytes**. Normally, no neutrophils are present; however bacterial meningitis causes neutrophils to appear in the cerebrospinal fluid. A traumatic lumbar puncture is a common cause of erythrocytes appearing in the CSF.

3) **E**
The systemized nomenclature of medicine, SNOMED, is a multiaxial classification system that codes according to the following definitions:
T = Topography
M = Morphology
D = Disease/diagnosis
E = Aetiology (e.g. infectious and parasitic disease)
F = Function (signs and symptoms)

4) **D**
Papillary renal cell (adeno)carcinoma has a better prognosis than clear cell renal cell carcinoma of equivalent grade/stage. Collecting duct carcinoma has a notoriously poor prognosis.
Oncocytoma and angiomyolipoma are benign tumours.

5) **A**
A widely used definition of **clinical governance** is: "A framework through which NHS organisations are accountable for continually improving the quality of their services and safeguarding high standards of care by creating an environment in which excellence in clinical care will flourish."

6) **B**
The glomus tumour is derived from the modified **smooth muscle cells** of the glomus body; the latter is a specialized arteriovenous anastomosis involved in thermoregulation.

7) **D**
The text describes sun-damaged skin with a non-invasive melanocytic lesion. Classically a **dysplastic naevus** is greater than 4mm in diameter, has a lentiginous proliferation of melanocytes, shows random cytological atypia, Rete ridge elongation and fusion and lamellar fibroplasia.

8) **D**
The Gleason grading system describes a grade 3 tumour as showing an infiltrative border, marked stromal invasion, **angular glands** with variation in shape, variably packed irregular masses with basophilic cytoplasm. Although grade 3 can show cribriform foci, **cribriform** architecture is more commonly a feature of grade 4

tumours. Hence the majority of this tumour is grade 3 with a minority of the tumour being grade 4; **3 + 4**.

9) **B**

A death should be referred to a coroner if:

The cause of death is unknown.

The death was violent, unnatural or suspicious. This would include suicide, homicide, road traffic accidents, infant deaths or iatrogenic deaths.

The death may be due to an accident. This includes workplace, domestic and road traffic accidents.

The death may be due to self-neglect or neglect by others. **This may include death in nursing homes** and those sectioned under the Mental Health Act.

The death may be due to an industrial disease related to the deceased's employment.

The death may be due to an abortion.

The death may be related to an operation.

The death may be related to alcohol, poisoning or drug use.

There is a death in custody.

A death certificate can be issued without referral to a coroner if:
- The patient was seen alive by the doctor within the previous 2 weeks.
- The doctor is satisfied that the death was due to natural causes.
- The doctor is reasonably sure of the cause.
- None of the above indications for coroner's referral are met.

Scenario **A** is unlikely to require a referral to a coroner because the patient was seen by his GP within the two week window and it is probable that the death was related to exertion. Hence it is likely to be a natural death. Scenario **B** requires referral to a coroner because it is a death in a nursing home (in addition it is common for these patients **not** to have been seen by a doctor within two weeks of death). Scenario **C** is unlikely to require referral because of the known history of lung cancer and chemotherapy. There is no indication in the history that the death at home was suspicious.

Scenarios **D and E** are descriptions of natural causes of death due to infectious disease.

10) **D**

The tumour is most likely to be a primary renal cell carcinoma; invasion of the hilar veins is defined as grade **T3b**.

11) **C**

Classically, a **pleomorphic adenoma** is a partially encapsulated mass with a chondromyxoid stroma with ducts and dark staining epithelial/myoepithelial cells. Polymorphous low grade adenocarcinoma contains monotonous cells with eosinophilic cytoplasm with variable architecture and characteristic targetoid lesions. Warthin's tumours are encapsulated benign lesions composed of double layers of eosinophilic epithelial cells adjacent to a dense lymphoid stroma. Adenoid cystic carcinoma is an infiltrative lesion composed of small dark cells with scanty cytoplasm arranged in tubular, solid or cribriform patterns with abundant hyaline material between the groups of tumour cells. Squamous cell carcinomas have a single component of cells with abundant pale eosinophilic cytoplasm and variable nuclear pleomorphism. Mitotic activity is usually evident and keratin is present in the well differentiated examples.

12) A

Toxoplasma infection can affect lymph nodes to cause a non-caseating lymphadenitis; interfollicular clusters of monocytoid B cells with abundant clear cytoplasm are characteristic of this process. TB infection would typically produce caseating granulomata. Mucor is a fungus that affects the immunosuppressed and morphologically shows hyphae that branch at 90 degrees. EBV causes large atypical lymphocytes to appear in the peripheral blood and the lymph nodes. HIV infected individuals may develop a persistent generalized lymphadenopathy with a subset developing a B cell lymphoma.

13) D

A biopsy of the minor salivary glands of the lip is the gold standard biopsy used to diagnose Sjögren's syndrome. However a biopsy of the **parotid gland** is also used to diagnose the syndrome.

14) C

Malignant mesothelioma can occur in people working in the building industry because of their exposure to asbestos fibres, is exacerbated by smoking and can present as chest pain. The cough might have beeen due to concurrent chronic bronchitis secondary to smoking. Emphysema is more likely to present as dyspnoea than as a cough; it is unlikely to present with chest pain. Asbestosis usually presents with progressive dyspnoea. Lung adenocarcinoma is a reasonable differential diagnosis however, the chest pain symptom and history of work in the building industry favour mesothelioma. A fever and productive cough would be expected in an established pneumonia.

15) B

Embryonal carcinoma occurs in a pure form but also as a secondary component in mixed germ cell tumours. The tumour cells are undifferentiated epithelial cells with abundant clear to granular cytoplasm. The cells can be arranged in a variety of architectures including pseudoglandular. CD30 positivity is common. Choriocarcinomas are composed of syncytiotrophoblasts and cytotrophoblasts in a haemorrhagic and necrotic background and are hCG positive. Yolk sac tumours have variable and mixed morphology but often contain Schiller-Duval bodies and frequently show AFP positivity. Leydig cell tumours consist of medium to large polygonal cells with abundant eosinophilc cytoplasm and distinct cell borders. Reinke crystals are frequently present and the tumours show positivity to steroid hormones. Intratubular germ cell neoplasia refers to dysplasia within tubules.

16) A

In an unscreened patient the gradual onset of the symptoms may mean that the diagnosis of familial adenomatous polyposis (FAP) may not be made until the patient is in their late twenties or thirties. The germline mutation is in the **APC** gene. p53 mutations occur in several cancers and are considered to be gateway mutations however they have no role in FAP. The Rb gene is another tumour suppressor gene; it predisposes the individual to retinoblastomas and osteosarcomas. HNPCC gene carriers do not have florid polyposis and the vast majority show microsatellite instability. C-myc is not associated with colorectal tumours.

17) **A**
The text describes **ductal carcinoma in situ**. No features of lobular, mucinous, medullary or papillary tumours are described.

18) **E**
The original name for the HER2 gene was **c-erb-B2**.

19) **D**
Although revalidation is mandated by the government, the collection of CPD points is organised by the **Royal College of Pathologists**.

20) **D**
Fine needle aspiration cytology classifies the results as:
C1 - Inadequate
C2 - Benign
C3 - Equivocal
C4 - Suspicious
C5 - Malignant

21) **C**
For **diffuse axonal injury** to occur a loss of consciousness must be part of the clinical picture. Indentation in the uncal region is a classical feature of transtentorial herniation and Duret's haemorrhage is a feature of progression of transtentorial herniation. The cerebral oedema and papilloedema require time to occur but are not excluded by the described timecourse.

22) **E**
The young woman has the signs and symptoms of **amniotic fluid embolism.** The long labour predisposes to trauma and access of amniotic fluid to the maternal circulation. Classically, the fluid contains fetal squames. Although pulmonary embolism can cause sudden maternal death, the context of a difficult labour, seizures and lack of leg swelling favour an amniotic fluid embolism. A phaeochromocytoma has acute signs of excessive catecholamine secretion. Cervical tear and haemorrhage is associated with visible bleeding and shock. Pre-eclampsia has a gradual onset and the increased blood pressure is often associated with CNS signs.

23) **C**
The child in question is suffering from meningitis. Worldwide the most common cause in an infant is **Haemophilus influenza**. Since the 1990s immunisation programs in the West have diminished the prevalence of Haemophilus Influenza, making Neisseria Meningitidis and Streptococcus pneumoniae more prevalent. Staphylococcus aureus and Streptococcus faecalis are uncommon causes of meningitis.

24) **B**
The description is consistent with malakoplakia. The areas of calcification are **Michaelis-Guttmann** bodies and occur as a result of the histiocytes inability to break

down the infecting bacteria. Psammoma bodies are extracellular only. Dystrophic calcification occurs as a focal event in areas of necrosis and is necessarily extracellular. Metastatic calcification occurs as a result of systemic increase in circulating calcium secondary to a metabolic inbalance. The calcification may be intracellular or extracellular and occurs as amorphous deposits. No parasitic morphology is described.

25) **B**

Classically immunosuppressed patients with HIV are prone to CNS infection by **cytomegalovirus**. This virus is reactivated in the immunosuppressed and can cause an encephalitis. Coccidiomycosis is a fungal infection that preferentially affects the lungs. Staphylococcus aureus is a skin commensal and usually acts on impaired skin. Streptococcal infections are common in children and can cause a range of pathologies, however they do not commonly cause infections in the adult brain. In immunosuppressed transplant individuals, EBV can cause a post transplant lymphoproliferative disorder.

26) **C**

The question describes **multiple endocrine neoplasia type 2A**. The mutant gene locus is RET and predisposes to parathyroid hyperplasia, phaeochromocytoma, C-cell hyperplasia of the thyroid and medullary carcinoma of the thyroid. Gardner's syndrome is related to classical familial adenomatous polyposis in producing similar intestinal polyps. Marfan's syndrome is a connective tissue disorder that occurs because of defective fibrillin-1. Tuberous sclerosis causes formation of hamartomas and benign neoplasms throughout the body – however the endocrine network is not a common site for these lesions. Von Recklinghausen's disease is an alternative name for neurofibromatosis type 1.

27) **D**

Assuming that there are no significant findings at autopsy then the findings are most consistent with **sudden infant death syndrome**. Pneumonia usually has some or all of the following signs and symptoms; fever, cough, breathlessness or lethargy. Croup tends to exhibit a night time onset of a barking cough, harsh stridor and hoarseness. A preceding fever is common. Meningitis can present in infants with fever, poor feeding, vomiting, irritability, lethargy, drowsiness, seizures, reduced consciousness or coma. Children that can talk may present with headache, neck stiffness and photophobia. Without witnessed recurrent seizures or a family history of a seizure disorder it is very unlikely that epilepsy would be a sudden cause of death in an infant.

28) **A**

As a general rule, during an autopsy technicians are most concerned about a transmitted infection on opening the **thoracic cavity**. Tuberculosis is the biggest concern here. Immigrants, asylum seekers, and refugees from countries where tuberculosis is endemic often relocate to countries where the incidence of tuberculosis is low. A significant percentage of such persons will develop renal failure requiring renal replacement therapy. Such chronic renal failure impairs immune function and is associated with an increased incidence of TB. *Among patients with chronic renal failure requiring renal replacement therapy, rates of TB 10 to 25 times greater than* *those in the general population* have been reported in North America and Europe. This is equivalent to incidence rates of approximately 250 cases per 100,000 per year.

Opening the skull can create an aerosol that includes bone fragments and bone marrow; in principle blood borne diseases can be transmitted. **E** is also a reasonable answer.

This is a difficult question and was included in the exam set to reinforce the fact that questions included in the MRCPath Part 1 examination are not always included in the final scoring. Questions that have two or more plausible answers have been excluded in the past and it is the responsibility of a prudent candidate not to waste too much time on such questions.

29) B

Necrotising enterocolitis occurs in preterm infants in the first few weeks of life and is thought to be due to a combination of ischaemia of the bowel wall and infection from organisms colonising the bowel. Bloody diarrhoea is often observed. Although Wilms tumour can cause abdominal distention, bloody diarrhoea is not a feature of this disease. Crohn's disease can cause bloody diarrhoea but does not usually cause vomiting. In addition, the marked changes to the bowel are unlikely to have occurred in such a young child as a result of Crohn's disease. Diverticulosis and hence diverticulitis is a disease of the middle aged and elderly.

30) A

The remit of the Coroner's autopsy is to ascertain the cause of death (essentially whether the cause of death was natural or unnatural). It seems likely that in this autopsy the cause of death is the head trauma. The investigation of the hilar lymphadenopathy is unlikely to yield any findings that will contribute to the cause of death, hence permission from the closest relative (**next of kin**) must be sought before proceeding further.

31) D

The description is most consistent with a **leiomyoma**. Leiomyosarcomas are rarer than leiomyomas, have poorly demarcated edges and frequently show zones of haemorrhage and necrosis. Endometrial stromal tumours can be localized but tend to widely permeate the myometrium and are usually softer than leiomyomas. Endometriosis usually presents as a haemorrhagic mass/chocolate cyst if visible macroscopically. Thecomas are typically ovarian tumours with yellow solid cut surfaces; cystic change, haemorrhage and necrosis are possible.

32) B

The current national cervical screening recommends that under normal circumstances, smears should be kept for **10 years**.

33) A

The average age of first onset of colorectal malignancy in **HNPCC** (Hereditary non-polyposis coli) is 45; individuals with HNPCC have an increased risk of sporadic colorectal cancers but do not suffer from florid polyposis coli. FAP (familial adenomatous polyposis) sufferers first develop >100 colorectal polyps in their second decade and often have invasive tumours in their third decade. The occurrence of sebaceous gland tumours together with an HNPCC type internal malignancy is designated as the Muir-Torre syndrome. Marfan's syndrome is a connective tissue disease. Gardner's syndrome is an FAP-like disorder which also causes multiple osteomas, epidermal cysts and fibromatoses.

34) B

The biggest risk factor for CIN (cervical intraepithelial neoplasia) is the sexually transmitted virus, human papillomavirus (HPV). A **barrier method** of contraception can physically inhibit infection. The advice to refrain from sexual activity, although ultimately successful if followed, is unlikely to elicit compliance. Smoking is a lesser risk factor for CIN. Alcohol use is not a directly contributory risk factor. Colposcopy is a treatment, rather than a screening, modality.

35) D

Arrthymogenic cardiomyopathy demonstrates fibroadipose infiltration of the myocardium and aneurysms of the right free wall. There is no history of alcohol abuse or any reliable diagnostic means of distinguishing this from other dilated cardiomyopathies. Dilated cardiomyopathies usually have a suggestive medical history and mutations in cytoskeletal and nuclear transport proteins. A hypertrophic cardiomyopathic heart shows left ventricular hypertrophy, outflow obstruction and septal impact lesions. Restrictive cardiomyopathy can be associated with a macroscopically normal heart but microscopically extensive fibrosis is present.

36) A

Informed consent requires effective communication; an **interpreter** is needed. Otherwise no procedure can be carried out.

37) A

The definition of emphysema describes the dilation of airspaces **distal to the terminal bronchiole**. **C** is the definition of chronic bronchitis. **D** is the definition of asthma.

38) C

Spina bifida is a congenital human abnormality; these abnormalities have genetic, environmental and undefined contributory factors and are hence **multifactorial**.

39) B

A synovial fluid specimen can be examined unstained under **plane polarized light microscopy** after centrifugation; there is no requirement for an alcohol wash. (A synovial biopsy would require washing and fixation in alcohol). Urate crystals are water-soluble and so must not be washed in saline. Dark field microscopy is used in relation to syphilis. There is no indication to decalcify monosodium urate crystals.

40) E

Angiomyolipomas express melanocytic markers and have the classic triphasic morphology of mature adipocytes, irregular benign vessels and smooth muscle. A malignant melanoma would usually possess architectural and cytological features of malignancy. A renal cell carcinoma would usually be monomorphic and classically has a clear cell character. An angiolipoma has only vascular and adipocytic components. A myolipoma is composed of short fascicles of smooth muscle with mature adipocytes.

41) **D**

Cushing's syndrome is frequently caused by adrenocortical adenomas; these benign lesions are usually less than 60g in mass. Tumours **>100g** are at risk of being malignancies with an associated poorer prognosis. The other factors are histological factors that are not described for this tumour, however more than 5 mitoses per 10 high power fields (hpf) and less than 25% clear cell morphology, are associated with malignancy. Cushing's syndrome implies a functional neoplasm, producing ACTH. This is more likely in benign lesions.

42) **A**

The first check should be on the glass slide to confirm **appropriate labelling**.

43) **B**

Granulosa cell tumours usually have solid and cystic areas and are yellow in colour. Granulosa cell tumours are positive to S100, CD99, **inhibin**, SMA, calretinin and cytokeratin. CA125 is positive in ovarian serous adenoma/adenocarcinomas. CEA is a useful stain for carcinomas/adenocarcinomas. OM-1 is positive in ovarian serous cystadenocarcinomas and endometrioid carcinomas. Alphafetoprotein is positive in germ cell tumours.

44) **A**

Gastrointestinal stromal tumours are CD117 positive and CD34 positive with a minority being positive to desmin; they are often located in the stomach or small intestine. Neither leiomyomas nor leiomyosarcomas are usually CD117 positive. Although lymphomas can occur in the small intestine the immunohistochemical profile would be positive for lymphocytic markers and negative for CD117, CD34 and SMA. Carcinoids have neuroendocrine morphology and immunohistochemical profiles.

45) **C**

Although **aspiration pneumonia** is difficult to diagnose reliably at autopsy, the combination of alcohol abuse and food material in the air passages allows a reasonable inference to be drawn. No description of consolidation or a purulent exudate is given in the question stem – hence bronchopneumonia and lobar pneumonia are unlikely. One would expect CNS signs with a cardiovascular medical history for the diagnosis of a stroke. Without a medical history of arrhythmia, this diagnosis is usually made by the exclusion of other pathologies.

46) **C**

HNPCC occurs because of underlying mismatch repair gene mutations causing microsatellite instability; choices A, B, D and E relate to microsatellite instability and the human genes that are important in the process. 5 different genes have been identified as involved in HNPCC pathogenesis; hMSH2, hMLH1, hPMS1, hPMS2 and hMSH6. **V-sis** has no role in microsatellite instability.

47) **E**

Minimal change nephropathy (minimal change disease) is a classical cause of nephrotic syndrome in children and can be seen as fusion of epithelial foot processes (podocytes) under electron microscopy. No changes are visible under light microscopy. Membranoproliferative glomerulonephritis occurs in adolescents and young adults and can show mesangial proliferation and membrane thickening under light microscopy. Focal segmental sclerosis shows some glomeruli that show sclerotic and hyalinized areas. Only class I lupus nephritis shows normal light microscopy appearances; this is unlikely to cause signs and symptoms of renal failure in a child. IgA nephropathy is well recognized as a histological mimic of several glomerular diseases, however it tends to be an adult disease.

48) B

A classical **giant cell tumour of the bone** is histologically benign and is prone to local recurrence. It sometimes metastasizes but it is not possible to identify the tumours that will metastasize. A giant cell tumour of the tendon sheath is usually located in the hands of middle-aged women and does not metastasize. Leiomyosarcomas and rhabdomyosarcomas are histologically malignant spindle cell neoplasms. A leiomyoma is a histologically benign lesion.

49) B

The morphology of the cells is consistent with either a adenocarcinoma or a mesothelioma; **adenocarcinomas** are Berep4 and AE1AE3 positive whereas a mesothelioma would usually be calretinin positive. The morphological descriptions and immunocytochemical profiles are inconsistent with lymphoma, small cell carcinoma or squamous cell carcinoma.

50) C

The individual has a social history that would predispose to an infection that is blood borne and appears to cause predominantly hepatic disease; **hepatitis C** is the most likely diagnosis. Although alcoholic disease is also possible in an individual with a tendency to addiction, this would have caused systemic illness and is unlikely to have occurred in such a young individual. There is no mention of alcohol abuse in the question stem. HIV, EBV and TB would have caused significant systemic (non-hepatic) symptoms.

51) D

Classically **Burkitt's lymphoma** is CD10, CD20, CD79a and c-myc positive. It is made up of monomorphic medium sized cells with a high mitotic rate.

52) B

Classically **chronic lymphocytic leukaemia** is CD5, CD23, CD43 and CD79a positive and CD10 negative.

53) H

Follicular lymphoma is a B cell lymphoma that is CD20 and CD79a positive. It expresses CD10 in around 65-70% of the cases. The architectural description is consistent with follicular lymphoma.

54) C

This B cell lymphoma is cyclin D1 positive and the description is consistent with **mantle cell lymphoma**.

55) E

This B cell lymphoma is CD10, CD20, CD79a and Bcl2 positive. The description is consistent with a **diffuse large B cell lymphoma**. CD5 may be positive or negative.

56) A
This vesiculobullous skin disease is characterized by subepidermal blistering with the blisters containing eosinophils and is occurring in a 70 year old person. This is a classical description of **bullous pemphigoid**.

57) E
Dermatitis artefacta occurs more commonly in women then men, frequently during their teens or young adulthood. Lesions are often bizarrely shaped with irregular outlines. Lesions do not evolve gradually but emerge almost overnight without any prior signs or symptoms. The lesions are usually found on sites that are readily accessible to the patient's hands, e.g. face, hands, arms or legs. (Note that the question stem makes no reference to histopathological findings.)

58) B
The deposition of IgA under immunofluorescence, together with subepidermal blisters containing neutrophils, is consistent with **dermatitis herpetiformis**.

59) H
Hailey-Hailey disease is recognized by its crumbling brick wall appearance.

60) L
This subepidermal blistering disease, with negligible cellular content, has deposits of IgG and C3. This is a description of **epidermolysis bullosa acquisita**.

61) C
The age of the patient, the diffuse cortical atrophy and longstanding dementia are consistent with **Alzheimer's disease**.

62) A
The festinant gait and depigmented substantia nigra are consistent with **Parkinson's disease**.

63) D
This is a description of **multi-infarct dementia**.

64) E
A rapidly progressing dementia in a young man with spongiform change in the cerebellum is likely to be **Creutzfeldt-Jacob disease**.

65) B
This is a description of **Huntingdon's chorea** – the movement disorder with atrophy in the caudate nucleus, putamen and globus pallidus are consistent with Huntingdon's chorea.

66) G

Although there may be a background Hashimoto's thyroiditis the presence of CD20 and CD79a positive cells in the thyroid are consistent with an occult **B cell lymphoma**.

67) **E**

This a description of a papillary carcinoma of the thyroid – further features would be nuclear inclusions and chewing gum colloid. The follicular architecture points towards **follicular variant of papillary carcinoma**.

68) **A**

The abundant thick and thin and colloid with bland follicles is suggestive of **nodular colloid goitre** (multinodular goitre). Follicles of varying size, background macrophages and bare nuclei are also frequently present.

69) **H**

A thyroid aspirate with bizarre and large cells with a background of tumour diathesis is suggestive of **anaplastic thyroid carcinoma.**

70) **B**

The abundance of macrophages and scarcity of follicular cells is consistent with a **thyroid cyst**.

71) **C**

The infiltration of the right ventricle by fibroadipose tissue is characteristic of **arrhythmogenic cardiomyopathy**. Aneurysms of the right ventricular wall are pathognomonic.

72) **I**

Myocyte disarray is characteristic of **hypertrophic obstructive cardiomyopathy**. Myocyte hypertrophy, outflow obstruction and impact lesions are present in typical cases.

73) **B**

Sudden death in **ischaemic heart disease** requires at least one of the major coronary arteries to have pinpoint atheromatous occlusion.

74) **L**

The description is of a traumatically induced catastrophic arrhythmia. **Commotio cordis** is a cause of sudden death due to trauma to the chest overlying the heart.

75) **M**

The implied age of this woman and the small brown heart are in keeping with **senile myocardium**.

76) **A**

The t(14:18)(q32;q21) translocation occurs in **follicular lymphoma**.

77) **F**

Alpha-1 antitrypsin deficiency is associated with a **PiZZ** genotype.

78) **B**

HNPCC is caused by a failure of mismatch repair; **hMLH1** is the human homologue of a mismatch repair enzyme.

79) J
Osteosarcoma is one of the tumours that is associated with mutations in the Retinoblastoma (**Rb**) gene.

80) C
Ewing's sarcoma is associated with the **t(11;22)(q24;12)** translocation.

81) A
CIN1 produces smears that contain cells with mild dyskaryosis.

82) D
Background tumour diathesis and atypical squamous cells are consistent with **squamous cell carcinoma**.

83) L
This is a low grade squamous intraepithelial lesion of exophytic architecture with koilocytosis and is entirely consistent with **condyloma acuminatum**.

84) F
The description is of **high grade (intestinal type) CGIN**. High grade CGIN shows obvious hyperchromasia, granular chromatin, nuclear stratification, increased numbers of normal and abnormal mitoses, apoptotic bodies, epithelial budding and intraluminal projections. Morphological subtypes include endocervical, endometrioid, intestinal, serous, clear cell and adenosquamous.

85) C
The description of this lesion is consistent with endometrial glands, stroma and blood and is located in the peritoneum. This is **endometriosis**.

86) F
This five year old child has symptoms of pneumonia, the commonest cause in this age range is **Haemophilus influenza**. The implication of the child's social background is that immunization may not have occurred.

87) K
A Lancefield group A **beta haemolytic streptococcal infection** that is untreated can cause a sore throat and post infective rheumatic fever. Migratory polyarthritis, carditis, subcutaneous nodules, erythema marginatum and Sydenham's chorea are major manifestations of rheumatic fever.

88) D
Until recently the commonest cause of neonatal meningitis was an E. coli infection. Now the commonest cause in the Western world is a **group B streptococcus** infection.

89) I

E. coli can cause haemolytic uraemic syndrome. This is a triad of acute renal failure, thrombocytopaenia and microangiopathic haemolytic anaemia.

90) **B**
Toxoplasma infection in the neonate (congenital toxoplasmosis) causes the classical triad of hydrocephalus, chorioretinitis and intracranial calcifications.

91) **E**
The **epiglottis** is at the superior boundary of the larynx.

92) **I**
The **middle turbinate** is a bony structure in the nasopharynx.

93) **L**
The submental lymph nodes are defined as cervical **level 1 nodes** (of a neck dissection).

94) **F**
The question describes a **nasal polyp**. The mucosa is oedematous and may contain hyperplastic or cystic mucous glands. It may be infiltrated by a range of inflammatory cells, including eosinophils.

95) **H**
The description is consistent with early invasion by a **squamous cell carcinoma**.

96) **A**
Type 1 collagen forms the major part of the extracellular matrix in bone:
Type 1 collagen is present in skin, bone and tendons.
Type 2 collagen is present in cartilage and vitreous humour.
Type 3 collagen is present in blood vessels, uterus and skin.
Type 4 collagen is present in basement membranes.
Type 5 collagen is present in interstitial tissues and blood vessels.
Type 6 collagen is present in interstitial tissues.

97) **H**
The **epiphyseal cartilage** is part of the growth plate, with ossification centres being on either side of the growth plate.

98) **J**
The marrow of haemopoietic long bones contains trabecula of mature bone that is **cancellous** bone.

99) **K**
A **sequestrum** is a piece of dead bone that has become separated within a cavity usually as a result of osteomyelitis-induced necrosis. Involucrum is new bone formed beneath the elevated periosteum that surrounds the sequestrum.

100) **N**
Paget's disease causes disorganized bone growth and resorption, resulting in mechanically weak bones that are prone to chalk stick fractures.

101) C

Hirschprung's disease is a congenital disorder in which ganglion cells are absent due to failure of migration of neuroblasts from the vagus nerve. The disease affects a varying length of large intestine extending proximally from the anus.

102) F

A **Wilms** tumour classically has blastematous, stromal and epithelial components.

103) A

Homer-Wright rosettes are most usually seen in **neuroblastomas**.

104) K

The age of the patient, site of the tumour and radiological characteristics are consistent with an **osteosarcoma**.

105) D

A **synovial sarcoma** classically has an epithelial and spindle cell component. The question describes the monophasic variant. The spindle cells are CD99 positive in 90% of tumours. Although the epithelial cells are the cytokeratin positive cells in a biphasic synovial sarcoma, some of the spindle cells are usually also cytokeratin positive in a monophasic synovial sarcoma.

106) D

The key to this question is a consideration of the paraneoplastic syndromes. Carcinoid syndrome can be caused by bronchial carcinoid, pancreatic carcinoma or gastric carcinoma. However, **Cushing's syndrome** is caused by small cell carcinoma of the lung, pancreatic carcinoma or some neural tumours.

107) I

This is a clinical description of the **carcinoid syndrome**.

108) A

Individuals with **Barrett's oesophagus** have an increased risk of developing invasive adenocarcinoma because of this pre-neoplastic disorder.

109) F

An individual with longstanding **coeliac disease** is at risk of enteropathy associated T cell lymphoma of the small intestine.

110) K

Zollinger-Ellison syndrome occurs as a result of gastrinoma causing hypergastrinaemia that in turn increases gastric acid secretion and eventually the formation of multiple peptic ulcers. The ulcers are resistant to standard treatment modalities.

111) F

In this scenario the ingestion of the recreational drug ecstasy should be considered and **blood MDMA** should be investigated.

112) G

The young woman's response to the meal can be explained by an ingested allergen causing an anaphylactic response and subsequent myocardial infarction. **Blood mast cell tryptase** levels will be increased in this Type 1 hypersensitivity reaction.

113) B
Young epileptics often have difficulty complying with their medication regime. The impulsive personality is consistent with this. Measuring the **phenytoin** levels in the hair will confirm or exclude compliance with the antiepileptic medication.

114) E
In non-tumour related occupational deaths that involve lung pathology, a minimum of four blocks should be taken (upper lobe, middle lobe/middle zone, upper aspect of lower lobe and lung base. This is to facilitate determination of the extent, distribution and causation. However the optimal number of **tissue blocks is 15** (*RCPath Guidelines on Autopsy Practice*, Scenario 7: Industrial/Occupational-related lung disease deaths including asbestos).

115) A
The car exhaust gases include **carbon monoxide**. In an enclosed space carbon monoxide poisoning becomes a possibility.

116) K
Calretinin and thrombomodulin are mesothelial markers, and DPAS would tend to be positive in mucin-producing adenocarcinomas. These results support the diagnosis of a **mesothelioma**.

117) H
The description is consistent with **coal worker's pneumoconiosis**; interstitial fibrosis and coal dust nodules are present. The larger nodules of progressive massive fibrosis (complicated coal worker's pneumoconiosis) are not present.

118) L
Adult respiratory distress syndrome (diffuse alveolar damage) is characterized by focal hyaline membranes, type II pneumocyte proliferation, interstitial fibroblastic proliferation, an inflammatory infiltrate and thrombi in the small pulmonary arteries.

119) A
The question stem describes the histological appearances of an **asthma**tic airway.

120) J
Apical caseating granulomas are pathognomonic of **tuberculosis**.

121) H
The prominent eosinophilic nucleoli and macrophages containing brown pigment should raise the suspicion of **malignant melanoma**.

122) A
The **classical seminoma** is composed of sheets of uniform clear cells, with distinct cell membranes, a large central nucleus and prominent nucleoli. Mitoses are infrequent.

123) **J**

This epithelioid tumour is PLAP negative, suggesting that it is not a germ cell tumour. The abundant eosinophilic cytoplasm and CK14 positivity are consistent with a **squamous cell carcinoma**. The groin lymph nodes are a frequent site of metastasis for squamous cell carcinoma of the genitalia.

124) **D**

Yolk sac tumours can demonstrate a range of histological patterns but are usually alphafetoprotein positive and often contain Schiller-Duval bodies and eosinophilic hyaline material.

125) **B**

The three different mature tissue types in this testicular mass are suggestive of a mature teratoma (**teratoma differentiated**).

PAPER 2

1) A 70 year old man with a 20 year history of smoking suffers from a progressive deterioration in lung function and eventually dies of suspected pneumonia. The deceased had spent 30 years as a manual worker in Sheffield's heavy industries. A coroner's autopsy is requested. The findings include a white firm mass encasing both lungs. Which investigations would you not carry out?

(A) Two random tumour blocks for immunohistochemistry.
(B) Two random lung tumour blocks for histology.
✓ (C) Lung blocks from upper lobe(s), middle lobe/zone, upper aspect of lower lobe and lung base.
(D) A lung block from each of the areas of macroscopic consolidation.
(E) Blocks for histology from the heart weighing 570g.

2) A 40 year old woman complaining of severe intractable headaches and visual field loss, with a one week history, has a CT scan of her head. The CT report describes a poorly defined spherical lesion in the right frontal lobe, with an associated midline shift. A neuropathologist stated that the subsequent biopsy showed "…..neoplastic cells that merge with normal glial cells. Moderate nuclear atypia is noted. Numerous normal mitotic figures are present, with sparse atypical mitotic figures. No vessels or necrotic foci are observed……"
The report is consistent with which of the following diagnoses?

(A) Pilocytic astrocytoma
(B) Grade 1 astrocytoma
(C) Grade 2 astrocytoma
✓ (D) Grade 3 astrocytoma
(E) Grade 4 astrocytoma

3) A 25 year old medical student presents as an acute hospital admission with a 24 hour history of severe headache and drowsiness. On examination she is photophobic and mildly pyrexial. Three hours after admission a witnessed seizure occurs. No rash is noted. Investigations reveal no evidence of alcohol or recreational drugs in his blood. Spinal fluid is drawn that reveals 300 white blood cells per ml with lymphocytes predominating and 70 mg/dl of glucose in the CSF.
The findings are consistent with:

(A) Normal CSF
(B) Viral encephalitis
✓ (C) Viral meningitis
(D) Bacterial encephalitis
(E) Bacterial meningitis

4) A 46 year old woman presents with right lower quadrant pain of gradual onset over the last 6 months. She also complains of pain during intercourse and the sensation of bloatedness. She has a personal medical history of recurrent dysplastic naevi and chlamydial infection. She did not attend her last cervical screening examination. 5 years ago she underwent treatment for linitis plastica. On examination and

investigation no evidence of pelvic infection was found. Histological examination of her right ovary is most likely to reveal:

(A) Malignant melanoma
(B) Leiomyosarcoma
(C) Endometrial stromal tumour
(D) Lymphoma
✓ (E) Krukenberg tumour

5) A 65 year old manual worker with a six month history of weight loss, has a pleural effusion that is tapped and sent for cytopathological assessment. The cells are polygonal with mild pleomorphism and exist as large clumps and free cells, some with macronucleoli. Immunocytochemistry is performed and the cells are positive to EMA (membranous staining) and calretinin but negative to AE1AE3 and BerEp4. What is the most likely diagnosis?

✓ (A) Malignant mesothelioma
(B) Adenocarcinoma
(C) Squamous cell carcinoma infiltrating the pleura
(D) Lymphoma
(E) Small cell carcinoma

6) A forty year old woman presents with signs and symptoms of renal failure. After biopsy, light microscopy reveals 40 glomeruli of which 2 are globally sclerosed and 38 show mesangial proliferation. The glomerular membranes appear thickened and wire loops are evident. Immunodeposition studies reveal diffuse and significant quantities of IgA, IgM, IgG and C3. Electron microscopy demonstrates mesangial, subendothelial and subepithelial deposits.
What is the likely diagnosis?

(A) Membranoproliferative glomerulonephritis
(B) Focal segmental glomerulosclerosis
✓ (C) Lupus nephritis ?
(D) IgA disease
(E) Minimal change disease

7) You are asked to cover the cut-up of macroscopic specimens for the urology team at very short notice, because the consultant is on sick leave. Most of the specimens are straightforward and you deal with them quickly and efficiently. The last specimen of the session is a right nephrectomy. You note that there is a well circumscribed spherical and cream coloured and haemorrhagic mass in the upper pole, 60 mm in maximum extent. The tumour is focally firm. Which of the following blocks are you least likely to take?

(A) Perinephric lymph node
(B) Adherent adrenal gland
(C) Renal vein resection margin
(D) Perinephric fat
✓ (E) 2 tumour blocks

53

8) You are reporting a quantity of excision skin specimens from a local GP. The majority are melanocytic naevi or basal cell papillomas (seborrheic keratoses). However you have particular difficulty with one of the specimens. Histologically you recognize a dermally based lesion with numerous spindle cells that show moderate pleomorphism and mitotic activity. The lesion has an infiltrative edge. Which of the following would not be included in your differential diagnosis?

(A) Malignant melanoma
(B) Squamous cell carcinoma
(C) Dermatofibrosarcoma
√(D) Granular cell tumour
(E) Leiomyosarcoma

9) On histological examination, a prostate biopsy shows invasive malignant glandular cells organised predominantly as sheets of packed cells, with a lesser component of cribriform architecture. Which of the following is the most accurate Gleason grade:

(A) 2 + 2
√(B) 5 + 4
(C) 3 + 3
(D) 4 + 3
(E) 5 + 3

10) A sigmoidectomy specimen is received after macroscopic excision of a biopsy confirmed primary adenocarcinoma of the colon. Three tumour blocks, proximal and distal resection margins, palpable lymph nodes and background mucosa are sampled. On microscopic examination a moderately differentiated adenocarcinoma is identified that just extends through the muscularis propria. Extramural vascular invasion is seen. 7 lymph nodes are identified, with 4 containing adenocarcinoma. The apical node is tumour-free. The proximal and distal resection margins are tumour-free. Which of the following is most accurate?

√(A) pT3 pN2 pMX, Dukes C1
(B) pT2 pN2 pMX, Dukes C2
(C) pT3 pN2 pMX, Dukes C2
(D) pT2 pN2 pMX, Dukes C1
(E) pT2 pN1 pMX, Dukes C2

11) A hospital-based dermatologist submits an ellipse of skin for histopathological assessment. The biopsy shows features of a superficial acute inflammation. At high power light microscopy, you notice multinucleate cells with intranuclear eosinophilic inclusions. The affected keratinocytes show swelling and separation from the adjacent cells. What is the likely cause of the infection?

(A) Cytomegalovirus
√(B) Herpes Simplex
(C) Hepatitis A
(D) Tinea
(E) Sarcoptes scabiei

12) A hard-pressed dermatopathologist reports an excision biopsy as:
"This specimen shows a viral wart. Excision appears complete. There is no evidence of dysplasia or malignancy." 6 months later the patient is shown to have an adjacent superficial spreading malignant melanoma. All the previous slides are reviewed. The report above is then shown to be inaccurate, because a small collection of malignant melanocytes is identified on one of the slides. The subsequent investigation of the dermatopathologist's performance requires a categorical classification of the expression of concern, for the discrepancy. Which of the following is the most appropriate category?

(A) E
(B) C
✓(C) B
(D) A
(E) D

13) Which body is most directly involved in re-licensing of doctors?

(A) DOH
✓(B) GMC
(C) BMA
(D) RCPath
(E) CPA

14) A patient complaining of severe knee pain undergoes aspiration of synovial fluid and sampling of deposits. The rheumatologist submits the fluid to the pathology department. He suspects that the patient has pseudogout. Which of the following results will be reported if he is correct?

(A) Positively birefringent sodium monourate crystals.
✓(B) Positively birefringent sodium pyrophosphate crystals.
(C) Negatively birefringent sodium monourate crystals.
(D) Negatively birefringent sodium pyrophosphate crystals.
(E) Abundant neutrophils only.

15) A thyroid gland has been submitted for macroscopic examination, cut-up and reporting. Unfortunately the clinician has forgotten to write a clinical history on the request form. You note the thyroid gland is grossly enlarged and symmetrical. The surface is smooth. On sectioning the gland shows randomly located firm white specks, cysts ranging from 5-10 mm in maximum extent, focal haemorrhage and scattered fibrotic foci. Which of the following are you least likely to find on microscopic examination?

(A) Significant variation in follicle size.
(B) Cholesterol crystals.
(C) Dystrophic calcification.
(D) TTF-1 positive tissue.
✓(E) Ground glass nuclei and adjacent psammoma bodies.

16) In embryological terms, the human breast is derived from:

(A) Sebaceous tissue
(B) Eccrine tissue
✓ (C) Sweat gland
(D) Trichal tissue
✓ (E) Apocrine gland

17) A 60 year old man presents to his general practitioner with pain and swelling of his left testis. There is no history of trauma. On palpation the testis is enlarged and firm. A tumour is suspected. If the tumour is PLAP and AFP negative, which of the following is the most likely diagnosis?

(A) Teratoma
(B) Seminoma
✓ (C) Lymphoma
(D) Small cell carcinoma
(E) Embryonal carcinoma

18) Whilst using a histopathology reference book you notice the following description of a breast cancer:
"…a distinct form of duct carcinoma with cuboidal to columnar cells that have eosinophilic granular cytoplasm. The cells are arranged in morule-like clusters that have an "exfoliative appearance." There is an inside-out growth pattern. Micropapillae without a fibrovascular core are characteristic. Most cases have associated intraductal carcinoma. There is lymphatic invasion in at least 50% of cases……"
The description most accurately applies to:

(A) Primary mucinous carcinoma
✓ (B) Invasive micropapillary carcinoma
(C) Medullary carcinoma
(D) Phyllodes tumour
(E) Spindle cell breast cancer

19) A thyroid FNA specimen reveals abundant thick and thin colloid, numerous follicles, dissociated follicular cells, hyalinised stroma, degenerating erythrocytes and sparse macrophages. What is the best classification for these observations?

(A) Thy5
(B) Thy4
(C) Thy3
✓ (D) Thy2
(E) Thy1

20) A 30 year old man has a personal medical history of chronic pancreatitis. His father was diagnosed with parathyroid and pituitary adenomas. What genetic abnormalities is the patient likely to have inherited?

(A) NF1, NF2

(B) p16INK4A
(C) BRCA1, BRCA2
(D) RET
√ (E) MEN1

21) Which of the following is not a cardiac cause of sudden death in an adult?

(A) Bridged coronary artery
(B) Hypertrophic obstructive cardiomyopathy
√ (C) Mitral stenosis
(D) Ischaemic heart disease
(E) Wolff-Parkinson-White syndrome

22) Which of the following is not true of gastrointestinal stromal tumours (GISTs)?

√ (A) They do not occur in the oesophagus.
(B) There is usually an abnormality in the c-kit gene.
(C) They arise from interstitial cells of Cajal.
(D) Size is a prognostic factor.
(E) They can be treated with Gleevec (imatinib).

23) Which of the following breast signs and symptoms is mismatched with the pathological explanation?

(A) Nipple retraction; tethering by invasive carcinoma.
(B) Bloody discharge; duct papilloma.
(C) Peau d'orange; impaired lymphatic drainage.
(D) Mobile lump; fibroadenoma.
√ (E) Cyclical breast pain; malignancy.

24) Which of the following is not a type of consent used in medicine?

(A) General consent
(B) Specific consent
(C) Explicit consent
√ (D) Inferred consent
(E) Implicit consent

25) A patient with a recent diagnosis of stomach cancer is disconcerted because there is no family history of stomach cancer. During further discussion with the patient, which of the subsequently elicited pieces of information would have increased his chance of stomach cancer?

(A) Trimethoprim
(B) High socioeconomic status
(C) South American ancestry
√ (D) Blood group A
(E) Fried foods

26) Which of the following is not a feature of good medical practice?

(A) Protecting and promoting the health of patients and the public.
(B) Supporting patients in caring for themselves to improve and maintain their health.
✓(C) Maintaining familiarity with (his/her) BMA number.
(D) Working in partnership with patients.
(E) Treating colleagues fairly and with respect.

27) "Permanent dilatation of bronchi and bronchioles caused by the destruction of muscle and elastic tissue." This is the definition of:

(A) Chronic bronchitis
(B) Emphysema
(C) Pneumonia
✓(D) Bronchiectasis
(E) Bronchiolitis

28) _____ _____ is an opportunistic infection in immunocompromised HIV carriers. It occurs usually when the **CD4 cell count** is less than **200 per ml of blood**. The chest radiograph may be normal or show bilateral midzone shadowing. Diagnosis is by histology of bronchial washings or biopsies. Treat with co-trimoxazole; prophylaxis with co-trimoxazole. The missing words are:

(A) Candida albicans
✓(B) Pneumocystis carinii
(C) Mycoplasma pneumonia
(D) Legionella pneumophila
(E) Staphylococcus aureus

29) A right cheek mass from a female patient is biopsied. The histological findings are of a well defined tumour with epithelial and lymphoid components. The epithelium is bilayered; the basal layer demonstrates basaloid triangular cells whereas the luminal cells have a granular oncocytic cytoplasm. The architecture reveals cystic and papillary projections. Lymphoid follicles are present. This is a(n)

(A) Oncocytoma
(B) Adenoid cystic carcinoma
✓(C) Warthin's tumour
(D) Salivary gland carcinoma
(E) Squamous cell carcinoma

30) A pathology trainee dissects an ovary that contains a tumour. It is a recurrence of an ovarian tumour that was inhibin negative but CK8, 18, EMA and CEA positive. The tumour is most likely to be a:

(A) Thecoma
(B) Fibrothecoma
(C) Fibroma
(D) Sclerosing stromal tumour
✓(E) Brenner tumour

31) A 17 year old patient presents with a pigmented skin lesion on the forearm that has been present for 2 years. This elevated lesion has recently become darker and ulcerated. An excisional biopsy of the lesion revealed a dermal melanocytic cell proliferation. The melanocytes showed moderate pleomorphism, increased mitotic activity, atypical mitotic figures and prominent nucleoli. The melanocytes were organized into a single nodule based in the reticular dermis, with an infiltrative edge. No atypical melanocytes were found in the epidermis. No solar damage is present in the adjacent skin.
The lesion is most likely to be a(n):

(A) Dysplastic naevus
✓(B) Nodular sclerosing malignant melanoma
(C) Superficial spreading malignant melanoma
(D) Acral lentiginous malignant melanoma
(E) Lentigo maligna melanoma

32) Which of the following is not a risk factor for aortic dissection?

(A) Hypertension
(B) Atherosclerosis
(C) Marfan syndrome
(D) Aortic aneurysm
✓ (E) Ehlers-Dunlus syndrome

33) Which of the following clinical scenarios does not require referral of the autopsy to a coroner?

✓(A) A woman with a history of HIV infection has an episode of sepsis and dies.
(B) A 17 year old boy is found dead with an empty bottle of diazepam beside him.
(C) A 65 year old smoker dies on the operating table during a repair of a ruptured abdominal aortic aneurysm.
(D) A recently deceased male was a non-smoker who was suspected to have died of mesothelioma.
(E) A previously healthy patient delivers successfully by Caesarian section but dies two hours later.

34) Which of the following is a Class 3 infectious agent?

✓(A) Rabies
(B) Variola
(C) Escherichia coli
(D) Marburg virus
(E) Lassa fever

35) A breast biopsy is taken from a patient and sent to the pathology department. It is cut up and reported by a trainee pathologist. In the report it is stated that "sections show a focus of sparse ductal structures and periductal stromal oedema with a haematoxyphilic tint. No breast lobules are present. There is no evidence of ductal carcinoma in situ or invasive malignancy.........."
What should be the last line of the report?

(A) This is benign gynaecomastia. ✓

(B) This is benign fat necrosis.

(C) This is a benign duct adenoma.

(D) This is a phyllodes tumour.

(E) This is a radial scar.

36) Which of the following is a cause of sudden (cardiac) death?

(A) Mitral stenosis.

(B) Aortic regurgitation.

(C) Hyperplastic obstructive cardiomyopathy.

(D) Arrhythmogenic right ventricular cardiomyopathy. ✓

(E) Arterial myxoma.

37) A 60 year old man presents with insidious onset of weight loss, chest pain and cough. A chest radiograph reveals a well-defined solitary nodule in the periphery of the right middle lobe. Histological examination reveals glandular cells showing moderate nuclear and cytological atypia. The cells are organized in a monolayer that appears to follow alveoli walls. Which of the following is the most likely diagnosis?

(A) Atypical adenomatous hyperplasia.

(B) Bronchioloalveolar carcinoma. ✓

(C) Clear cell renal cell carcinoma.

(D) Squamous cell carcinoma

(E) Small cell carcinoma

38) A 25 year old woman presents with a persistent mass in her left forearm. Biopsy reveals an infiltrative monophasic tumour consisting of spindle cells with mild nuclear pleomorphism. A herringbone architecture is present. Sparse mast cells are seen. Foci of calcification are noted. The neoplasm is CD99 positive and AE1AE3 positive. Which of the following is the most likely diagnosis?

(A) Fibrosarcoma

(B) Synovial sarcoma ✓

(C) Carcinosarcoma

(D) Leiomyosarcoma

(E) Atypical fibroxanthoma

39) In Whipple's disease the jejunal lamina propria contains numerous:

(A) Eosinophils.

(B) Macrophages. ✓

(C) Mast cells.

(D) Plasma cells.

(E) Neutrophils.

40) The true vocal cords are lined by:

(A) Ciliated columnar epithelium.

(B) Non-ciliated columnar epithelium.

✓(C) Stratified squamous epithelium.
(D) Pseudostratified epithelium.
(E) Intestinal type mucosa.

41) The weight range for a normal placenta is:

(A) 100 – 350g
✓(B) 350 – 700g
(C) 700 – 900g
(D) 900 – 1100g
(E) 1100 – 1300g

42) Which of the following statements regarding cirrhosis of the liver, is not correct?

(A) New vascular channels are formed in the liver.
(B) Collagen is laid down in the space of Disse.
(C) Fenestrations in the sinusoids are obstructed.
(D) The perisinusoidal cells (Ito) change into myofibroblasts.
✓(E) The architecture is altered into degenerating nodules bounded by elastin.

43) A 65 year old woman presents at her general practitioner complaining of recent onset temporal pain and transient blurred vision in the eye on the same side. Which one of the following is the likely diagnosis?

(A) Wegener's granulomatosis
(B) Churg-Strauss disease
✓(C) Giant cell arteritis
(D) Kawasaki disease
(E) Takayasu arteritis

44) Which of the following proteins is deposited in the brain in Lewy body dementia?

(A) Tau protein
(B) Amyloid
✓(C) Alpha-synuclein
(D) A-beta protein
(E) Alphabeta-crystallin

45) Which of the following is the least significant prognostic factor in a carcinoid tumour located in the lungs?

✓(A) Cell density
(B) Lymphatic/vascular invasion
(C) Mitotic count
(D) Necrosis
(E) Size

46) A 7 year old child presents with a history of lethargy, painful bone swelling and lymphadenopathy. Biopsy of a lymph node demonstrates eosinophils, neutrophils, histiocytes and small lymphocytes. A population of grooved cells with

lobulated/indented nuclei, and moderate amounts of faintly eosinophilic cytoplasm, are present. These cells show increased mitotic activity. The cells also demonstrate:

(A) NK1C3 positivity
√ (B) Birkbeck granules
(C) CD2a positivity (usually CD1a pos)
(D) HMB45 positivity (usually S100 posit)
(E) CD15 positivity (usually negative)

47) Which of the following is the greatest risk factor for prematurity?

√ (A) Maternal smoking
(B) Herpes
(C) Toxoplasmosis
(D) Cytomegalovirus
(E) Rubella

48) You are a new Head of Department and one of your pathology consultants is very reluctant to engage with Continuing Professional Development (CPD). Which of the following arguments would be least successful in persuading him to take CPD seriously?

(A) CPD helps doctors to comply with employers' and professional bodies' requirements.
(B) CPD provides evidence for annual appraisal and revalidation.
(C) CPD contributes to clinical governance.
√ (D) CPD requires no extra work.
(E) CPD prepares doctors for new roles (e.g. managerial).

49) Which fixative would you use to fix a renal biopsy for electron microscopy?

(A) Bouin's solution
(B) Formalin
√ (C) Glutaraldehyde
(D) Silver stain
(E) Grocott's stain

50) Which of the following is important in the diagnostic test for Hirschprung's disease?

(A) DPAS
(B) ABDPAS
(C) Sirius red
(D) PAS
√ (E) Acetylcholinesterase

A) Desmoplastic small round cell tumour   F) Renal sarcoma
B) Rhabdomyosarcoma   G) Small cell osteosarcoma
C) Lymphoblastic lymphoma   H) Wilms tumour
D) Ewing's sarcoma   I) Neuroblastoma
E) Peripheral neuroectodermal tumour

Which of the tumours A-I best matches the description below?

51) A biopsy of a mass was taken from the abdominal cavity of a 9 year old female. This infiltrative monomorphic tumour of small round blue cells is organised into rosettes with fibrillary cores. The cells have little cytoplasm and speckled nuclei. Necrosis and calcification are evident. Sparse ganglion-like bodies are seen.

52) A 10 year old male presents with multifocal abdominal masses. The biopsy of this tumour shows two distinct phases – stromal and malignant cohesive epithelial cells. The tumour is positive to immunohistochemical challenge from, NSE, cytokeratins, vimentin, desmin and CD99. It is negative to challenge from LCA and MSA.

53) These infiltrative discohesive cells have scanty basophilic cytoplasm, round nuclei and delicate chromatin. The tumour cells are CD7, CD99, LCA and TdT positive.

54) A 19 year old man has a mass in the medullary cavity affecting the metaphyseal region of his right femur. Histology indicated two populations of cells; large viable cells and smaller necrotic cells. This tumour has the translocation t(11;22)(q24;q12).

55) This malignancy is MyoD1 and MSA positive.

A) Aortic valve
B) Mitral valve
C) Tricuspid valve
D) Pulmonary valve
E) Aortic (valve) stenosis
F) Aortic (valve) regurgitation

G) Mitral (valve) stenosis
H) Mitral (valve) regurgitation
I) Tricuspid (valve) stenosis
J) Tricuspid (valve) regurgitation
K) Pulmonary (valve) stenosis
L) Pulmonary (valve) regurgitation.

Match the descriptions below to the appropriate heart valve/heart valve pathology:

56) Marfan's syndrome predisposes to this recognised cause of adult sudden death.

57) The prevalence increases with age and produces an ejection systolic murmur.

58) Annular ring dilation can be the underlying cause of this collapsing pulse.

59) A common site of infective endocarditis in intravenous drug users.

60) Fallot's tetralogy directly affects this valve.

A) Pancreas   F) Ovary   K) Testis
B) Bladder   G) Liver   L) Kidney
C) Colon   H) Lung
D) Prostate   I) Smooth muscle
E) Brain   J) Skeletal muscle

Choose one of the above options to answer each of the following questions:

61) What is the probable primary site of a metastatic adenocarcinoma that is EMA, RCC and alphabetacrystallin positive and CK7 negative?

62) What is the probable primary site of a metastatic adenocarcinoma that is CK7 and TTF1 positive but CK20 negative?

63) What is the probable primary site of a metastatic adenocarcinoma that is CK7, CK19, Ca19.9 and CK20 positive?

64) What is the probable primary site of a metastatic malignancy that is CK7, CK8, CK18 and CK20 positive?

65) What is the probable primary site of a metastatic adenocarcinoma that is CK7, ER, and CA125 positive but CK20 negative?

A) Focal nodular hyperplasia.   F) Hepatitis C   K) Hepatitis A
B) Macrovesicular steatosis.   G) Primary sclerosing cholangitis
C) Microvesicular steatosis.   H) Primary biliary sclerosis
D) Alcoholic steatohepatitis   I) Haemochromatosis
E) Hepatitis B   J) Wilson's disease

Match the liver diseases to the descriptions below:

66) This lesion has an arteriographic centrifugal filling pattern and a central stellate scar.

67) The pathogenetic virus has a three week incubation period.

68) Mallory bodies and zone 3 liver cell ballooning are present.

69) The liver stains positively with Shikata-Orcein in this pre-cirrhotic disease.

70) Associated with ulcerative colitis. "Onion skinning" may be present.

A) CIN1   E) Reparative changes   I) CMV infection
B) CIN2   F) Herpes infection   J) CGIN
C) CIN3   G) Trichomonas infection   K) Endometrial carcinoma
D) Candidal infection   H) Squamous cell carcinoma   L) HPV infection.

Select the diagnosis (A-L) that best matches with descriptions of the cytological preparations below:

71) 35 year old woman. The smear is predominately superficial squamous cells. Some of the squamous cells are binucleate. Cells with equivocal nuclear enlargement and perinuclear halos are noted. Keratotic features are present.

72) 38 year old woman. Clusters of syncytial squamous cells containing neutrophil microabscesses are present. The squamous cells have prominent nucleoli but smooth nuclear membranes. No other abnormality is seen.

73) 43 year old woman. This preparation shows keratotic strap-shaped and tadpole-shaped squamous cells with high nuclear to cytoplasmic ratios. The background shows necrotic tissue and cell debris.

74) 32 year old woman. Normal squamous cells are present. Scattered rosettes showing feathering, with nuclei at the tips, are evident. There is an inflammatory background.

75) 53 year old woman. There are numerous septate hyphae throughout the preparation. Inflammatory cells are noted. Very sparse 3D clusters of cells with scalloped outlines, high nuclear to cytoplasmic ratios and cytoplasmic mucin are present.

Match the following genes/gene products/genetic changes to the disease states given:

A) t(11;22)(q24;q12)   E) Myosin   I) TNF alpha
B) mHTT protein   F) NRAS   J) t(10;20)(q23;q12)
C) Dystrophin   G) Factor IX   K) 16p13.3   (PKD1)
D) Anti-GBM COL4-A3   H) PiZZ   L) Anti-GBM FAL9-B3
   antigen   antigen

76) Ewing's sarcoma

77) Huntington's disease

78) Follicular thyroid carcinomas

79) Polycystic kidney disease

80) Goodpasture syndrome

A) PECAM-1   F) Chemokines   K) TNF
B) ICAM-1   G) Fibrin   L) IL-1
C) Integrin   H) Fibronectin   M) Proteoglycan
D) P-selectin   I) Macrophage   N) Nitric oxide
E) E-selectin   J) Neutrophil

Choose the most appropriate inflammatory components from the list above, for each of the descriptions below:

81) A phagocytic cell important in chronic inflammation.

82) Selectin important in neutrophil-endothelial adhesion.

83) A cell surface glycoprotein important in neutrophil transmigration.

84) A chemical mediator involved in vasodilation produced by endothelial cells.

85) A class of chemical mediators that attracts neutrophils to sites of injury.

A) Type I collagen   F) Hepatitis C   K) Micronodular cirrhosis
B) Type II collagen   G) Viral hepatitis   L) Macronodular cirrhosis
C) Type III collagen   H) Hepatitis A   M) Alcoholic liver disease
D) Type IV collagen   I) Alpha-1 antitrypsin deficiency
E) Wilson's disease   J) Primary hemochromatosis

Select the most appropriate choice A-M, which applies to the questions below.

86) Commonest cause of cirrhosis in the West.

87) This type of collagen is not present in a normal or cirrhotic liver.

88) A patient with this disease developed diabetes mellitus and slate gray skin.

89) This type of hepatitis is unlikely to cause cirrhosis.

90) A cirrhotic patient presents with bilateral focal green pigmentation of the corneal limbus.

A) Fibrin
B) Ca^{2+}
C) Intrinsic pathway
D) Extrinsic pathway
E) Phospholipid
F) Thrombin

G) Hageman factor
H) Factor VIII
I) Factor IX
J) Factor X
K) Factor XI
L) Factor XII

Select the appropriate component above with the coagulation scenarios below:

91) Classically, this coagulation pathway is triggered by tissue injury.

92) These molecules are cofactors for several steps in the coagulation cascades.

93) The total activity of this factor is diminished in Haemophilia A.

94) Plasmin causes proteolysis of this clotting factor.

95) This factor is aberrant in Christmas disease.

A) Churg-Strauss syndrome F) Polymyositis K) Systemic sclerosis
B) Wegener's granulomatosis G) Dermatomyositis L) Temporal arteritis
C) Rheumatoid disease H) Kawasaki disease M) Psoriasis
D) Amyloidosis I) Henoch-Schonlein purpura
E) Ankylosing spondylitis J) Systemic lupus erythematosus

Which of the diseases above applies to the scenarios below?

96) This patient's X-ray radiograph shows a "bamboo spine."

97) This disease causes a "full house" of immunological deposition in the glomerulus.

98) An 8 year old child presents with signs and symptoms of acute cardiac chest pain.

99) A 78 year old man presents with recent onset severe headaches and pain when combing his hair. An urgent ESR/CRP measurement is abnormally high.

100) An adult male patient complains of recent onset wheezing. The lung biopsy shows vasculitis and the blood shows eosinophilia.

A) Haemoglobin
B) Myoglobin
C) Spherocytosis
D) Alpha thalassemia
E) Elliptocytosis

F) Sickle cell disease
G) Pernicious anaemia
H) Aplastic anaemia
I) Beta thalassemia
J) Glucose-6-phosphate dehydrogenase deficiency

K) Folate deficiency
L) Hexokinase deficiency

Choose the answer (A-L) that matches each of the scenarios below:

101) A lecturer in organic chemistry presents with anaemia, thrombocytopaenia and neutropaenia.

102) A tetrameric oxygen carrying protein.

103) This disease can be caused by an autoimmune attack on intrinsic factor.

104) A 50 year old woman of Afro-caribbean ethnicity has signs and symptoms of nephrotic syndrome.

105) An oxygen binding skeletal muscle protein.

A) ARDS G) Asbestosis L) Farmer's lung
B) UIP H) Silicosis M) Berylliosis
C) NSIP I) Coal worker's pneumoconiosis
D) COP J) Progressive massive fibrosis
E) Caplan syndrome K) Stannosis

The above are chronic interstitial lung diseases, occupational pneumoconioses and hypersensitivity lung diseases. Which of these diseases match the scenarios below?

106) This lung biopsy of a 50 year old ex-miner with a 10 year history of interstitial lung disease demonstrates anthracotic fibrotic nodules 2-7cm in maximum extent.

107) A 45 year old woman has a two week history of pneumonia requiring admission to an Intensive Care Unit. She does not respond well to the intravenous antibiotics. A lung biopsy reveals numerous hyaline membranes.

108) A 70 year old ex-coal miner has longstanding arthritis that shows ulnar deviation. He reports recently worsening shortness of breath. Examination reveals a 0.5 x 0.5cm subcutaneous nodule at his right elbow.

109) This lung disease is characterized by a type III hypersensitivity reaction to inhaled biological material.

110) This fibrotic lung disease is characterised by the presence of amphibole fibres.

A) MEN1   F) Carcinoid tumour   K) Parathyroid adenoma
B) MEN2A   G) Insulinoma   L) Pinealoma
C) MEN2B   H) Lipoma
D) Somatostatinoma   I) Gastrinoma
E) Adrenocortical adenoma   J) Leiomyoma

Which of the above diseases is being described in each of the following scenarios?

111) A 45 year old patient has experienced multiple intractable peptic ulcers. She also has a recent history of hypercalcaemia. An MRI scan demonstrated well-defined masses, 2cm in diameter, in the posterior cervical triangle and anterior pituitary.

112) This patient complains of bouts of flushing. The patient's urine contains increased concentrations of 5-HT metabolites. Biopsy of a new lung mass reveals a neuroendocrine tumour of low mitotic activity.

113) This benign tumour can cause Zollinger-Ellison syndrome.

114) A patient with an enlarged thyroid undergoes a biopsy and a medullary thyroid carcinoma is diagnosed. Further investigations reveal a phaechromocytoma and hyperparathyroidism.

115) This patient, patient's father and grandfather suffered from multiple endocrine neoplasias. All three generations also complained of mucosal lesions identified as mucosal neuromas.

A) Coup injury   F) Hydrocephalus   K) Diffuse axonal injury
B) Contrecoup injury   G) Epilepsy   L) Subarachnoid haemorrhage
C) Lymphoma   H) Dementia pugilistica
D) Meningioma   I) Epidural haemorrhage
E) Astrocytoma   J) Subdural haemorrhage

Which of the above pathologies is most applicable to each of the descriptions below?

116) A 70 year old woman fell and suffered head trauma without loss of consciousness. Two months later she collapsed and lost consciousness. She had no other significant past medical history.

117) This complication of head trauma is usually caused by bleeding of the middle meningeal artery.

118) A 45 year old woman presented with this tumour that was believed to be a complication of prior head trauma.

119) A patient died after a frontal head injury in a road traffic accident. The autopsy reveals occipital lobe bleeding and bruising. What is the best description of this pattern of injury?

120) A posterior fossa tumour in the foramen of Luschka can cause this condition.

A) Fibroadenoma
B) Phyllodes tumour
C) Paget's disease
D) Fat necrosis
E) Tubular carcinoma

G) Nipple adenoma
H) Fibrocystic change
I) Micropapillary carcinoma
J) Medullary carcinoma
K) Ductal carcinoma in situ

L) Ductal carcinoma
M) Lobular carcinoma in situ

Which of the breast diseases applies to the scenarios below?

121) A 15 year old girl presents with a mobile well-defined mass 2 x 2cm in the upper outer quadrant of her left breast. A biopsy of the mass confirms a benign lesion.

122) A 60 year old woman presents with a sanguineous nipple discharge. Biopsy of a subcutaneous nodule reveals a focus of well-defined adenosis with associated sclerotic changes. The lesion is adjacent to the epidermis of the nipple.

123) Mammography of this 50 year old woman revealed foci of calcification. Biopsy and histological examination demonstrated ductules with highly atypical cells proliferating in a solid pattern. The nuclei were markedly pleomorphic, poorly polarized and had clumped chromatin. Comedo necrosis with amorphous microcalcification was present. No invasion was found.

124) If high grade, this fibroepithelial lesion has abundant and very cellular stroma with frequent normal and abnormal mitotic figures.

125) This well circumscribed carcinoma has a syncytial architecture arranged in large sheets, accounting for at least 75% of the tumour mass. No glandular structures are present and there is scanty stroma. However there is a prominent lymphoplasmacytic infiltrate.

ANSWERS

1) C	43) C	85) F
2) D	44) C	86) M
3) B	45) A	87) B
4) E	46) B	88) J
5) A	47) A	89) H
6) C	48) D	90) E
7) E	49) C	91) D
8) D	50) E	92) E
9) B	51) I	93) H
10) A	52) A	94) A
11) B	53) C	95) I
12) C	54) D	96) E
13) B	55) B	97) J
14) B	56) H	98) H
15) E	57) E	99) L
16) C	58) F	100) A
17) C	59) C	101) H
18) B	60) K	102) A
19) D	61) L	103) G
20) E	62) H	104) F
21) C	63) A	105) B
22) A	64) B	106) J
23) E	65) F	107) A
24) D	66) A	108) E
25) D	67) K	109) L
26) C	68) D	110) G
27) D	69) J	111) A
28) B	70) G	112) G
29) C	71) L	113) J
30) E	72) E	114) B
31) B	73) H	115) C
32) E	74) J	116) K
33) A	75) K	117) J
34) A	76) A	118) D
35) A	77) B	119) B
36) D	78) F	120) G
37) B	79) K	121) A
38) B	80) D	122) G
39) B	81) I	123) K
40) C	82) E	124) B
41) B	83) A	125) J
42) E	84) N	

Paper 2 - Answers

1) **C**

The deceased had a history of smoking and probable asbestosis exposure. He appeared to die of pneumonia. The post mortem findings are consistent with the presence of a mesothelioma. (A) and (B) would confirm the presence and type of mesothelioma. (D) would confirm the pneumonia and the extent to which it contributed to the demise. (E) would give an indication of whether poor cardiac function contributed to the demise. However the **blocks in** (C) are those recommended for non – neoplastic lung disease by the RCPath Autopsy guidelines. The presence of the mesothelioma and an exposure history makes this unnecessary.

2) **D**

The medical history is consistent with a brain tumour. The histological report implies a similar morphology between the tumour cells and the glial cells "that merge." Glial tumours have 4 grades:

Grade 1 – well differentiated astrocytoma.

Grade 2 – nuclear atypia present.

Grade 3 – mitotic activity present.

Grade 4 – Vessel proliferation/necrosis present.

The histopathological report states absence of necrosis and vascular proliferation – making this a **Grade 3 astrocytoma**.

3) **B**

Fever, headache, photophobia and seizures are common in **encephalitis**. Neck stiffness, sudden high fever, severe headache and altered mental status are common in meningitis. The clinical picture favours encephalitis over meningitis. The following classical findings in the cerebrospinal fluid aid in the diagnosis:

Summary of typical CSF findings

	Normal	Bacterial infection	Viral infection
Cells	0-5 wbc/ml	>1000 wbc/ml	<1000 wbc/ml
Polymorphs	0	Predominate	Early
Lymphocytes	5	Late	Predominate
Glucose (CSF)	40-80mg/dl	Decreased	Normal
CSF/plasma glucose ratio	66%	<40%	Normal
Protein	5-40 mg/dl	Increased	+/-
Culture	Negative	Positive	Negative
Gram	Negative	Positive	Negative

Our medical student has normal CSF glucose and less than1000 white blood cells per ml with the appearances dominated by lymphocytes. This picture is consistent with a **viral** infection.

4) **E**

The **Krukenberg** tumour is a metastatic poorly differentiated adenocarcinoma in the ovary; histological examination typically reveals signet ring cells. The primary site is the stomach or the breast.

The patient is the right age range for a Krukenberg tumour (perimenopausal or older). Abdominal pain, bloating and pain during intercourse are common presenting symptoms of Krukenberg tumours. A history of linitis plastica (diffuse gastric adenocarcinoma) is a typical precursor to the Krukenberg tumour.

All of the other tumours are possible but less likely than a Krukenberg; for example, the diagnosis of melanoma would require an unnoticed morphological change consistent with malignancy, followed by metastasis to the ovary. Furthermore, one would expect the other tumours to show some of the systemic signs of malignancy; Krukenberg tumours do not show systemic signs until late in the progression.

5) **A**

The presence of a pleural effusion, together with the offered clinical history, are consistent with a malignancy. The cytological examination confirms pleomorphism and significant atypia but no diagnostic features are described. The immunohistochemistry shows a positive response to calretinin, a reliable marker of mesothelial cells. The membranous staining of the cells after EMA challenge is strongly suggestive of **malignant mesothelioma**. A squamous cell carcinoma is likely to have been positive to AE1AE3 challenge and negative to calretinin. Small cell carcinoma is unlikely to show clumped cells or macronucleoli. Lymphoma cells are generally discohesive and are unlikely to be calretinin positive.

6) **C**

The key diagnostic feature is the "full house" of immunodeposition. This classically occurs in **lupus nephritis**, which can also have a wide range of histological appearances. None of the other answer choices produces this range of immunodeposits. In addition the clinical picture is consistent with a sufferer from lupus nephritis.

7) **E**

The macroscopic description is consistent with a renal cell carcinoma. Invasion of the perinephric fat would require the tumour to be staged at a minimum of pT3a. Invasion of the renal vein would require staging at a minimum of pT3b. Identifying tumour in lymph nodes or the adrenal gland is rare, but significant. The minimum dataset requires that the tumour is adequately sampled, particularly if there are firm areas in the mass. Taking **only 2 blocks** of this 60 mm tumour, is unlikely to be adequate. Such an approach is likely to necessitate several repeat visits to the specimen to gain adequate sampling.

8) **D**

Spindle cell lesions of the skin are a common diagnostic conundrum. The differential for these circumstances should include:

Dermatofibroma/dermatofibrosarcoma
Malignant melanoma
Squamous cell carcinoma
Leiomyoma/Leiomyosarcoma
Schwannoma/Neurofibroma
Atypical fibroxanthoma
Vascular tumours

The possibilities can be reduced by the judicious use of immunohistochemistry and by due attention to the clinical history. **Granular cell tumour** does not typically have spindle cell morphology and should not be on the differential diagnosis.

9) **B**
The question stem assumes that the malignant cells represent a primary prostatic malignancy. These adenocarcinoma cells are predominantly poorly differentiated because they are present as sheets of cells, with no mention of gland lumina. Thus they are grade **5** cells. The groups of cells with cribriform morphology are defined as grade **4** in the Gleason classification.

10) **A**
The T N M staging system for colorectal adenocarcinomas is:

For tumour location
Submucosa (pT1)
Muscularis propria (pT2)
Beyond muscularis propria (**pT3**)
Tumour cells have breached the peritoneal surface
or invaded adjacent organs (pT4)

Node involvement by tumour
pN1 1-3 nodes involved
pN2 4+ nodes involved

Metastases
M0 – No metastasis
M1 – Tumour in a distant organ
MX – Metastatic status unknown

Dukes Classification
Dukes A (Invasion less than the full thickness of the muscularis propria, nodes negative)
Dukes B (Growth beyond muscularis propria, nodes negative)
Dukes C1 (Nodes positive and apical node negative)
Dukes C2 (Apical node positive)

11) **B**
The biopsy specimen demonstrates an inflammatory dermatosis. The microscopic description implies ballooning degeneration of keratinocytes and secondary acantholysis. The described intranuclear eosinophilic inclusions are typical of a **herpes simplex virus** infection. A more complete description might have included the identification of ground glass chromatin.
Tinea (versicolor) is a fungal infection that produces a characteristic "sandwich sign" in the keratin (above and below the hyphae). Cytomegalovirus (CMV) inclusions are intranuclear as well, but are darker staining and usually have peri-nuclear clearing. Hepatitis A does not directly infect the skin. Sarcoptes scabiei is parasite that forms burrows in the stratum corneum.

12) **C**

The Royal College of Pathologists published a "Review of the categorisation of discrepancies in histopathology" in November 2008. This document describes the response to an expression of concern about a doctor's performance. The following categories are described:

Category (Expression of concern)	Description
A	**Inadequate dissection, sampling or macroscopic description.**
B	**Discrepancy in microscopy.** 1. A diagnosis which one is surprised to see from any pathologist (e.g. an *obvious* cancer reported as benign). 2. A diagnosis which is fairly clearly incorrect, but which one is not surprised to see a small percentage of pathologists suggesting (e.g. a moderately difficult diagnosis, or missing a small clump of malignant cells in an otherwise benign biopsy). 3. A diagnosis where inter-observer variation is known to be large (e.g. disagreements between two adjacent tumour grades, or any very difficult diagnosis).
C	**Discrepancy in clinical correlation.** This would represent a failure to answer the clinical question.
D	**Failure to seek a second opinion in an obviously difficult case.**
E	**Discrepancy in report** This would include typographical errors and internal inconsistencies or ambiguities in the report which should have been corrected before authorization.

Our unfortunate dermatopathologist has experienced a **discrepancy in microscopy –** category **B**. Specifically the category is **B2**, because a (small) number of pathologists might have made the same error.

13) **B**
The **GMC** is directly responsible for the re-licensing process.

14) **B**
Gouty tophi contain negatively birefringent sodium monourate crystals. Pseudogout produces **positively birefringent sodium pyrophosphate crystals**. Abundant neutrophils would be associated with septic arthritis – which does not routinely produce deposits.

15) **E**

An incidental finding of a **papillary carcinoma** of the thyroid is less likely than the primary diagnosis of multinodular goitre. A-C represent diagnostic features of multinodular goitre. TTF-1 is thyroid transcription factor-1; the thyroid will show positivity to this challenge.

16) **C**

In mammals, the breast develops from **modified sweat glands**. The skin and underlying mesenchymal tissue differentiate under the influence of paracrine and systemic hormones.

17) **C**

A primary tumour of the testis is more common and more likely than a metastasis. In this age range seminoma, teratoma and embryonal carcinoma are unlikely as they occur in younger individuals. Seminomas are classically PLAP positive and embryonal carcinomas are usually AFP positive. **Lymphoma** is a common testicular cancer in this patient's age range and is usually PLAP negative.

18) **B**

The description given in the question stem is of invasive **micropapillary carcinoma**. The "inside-out" nature of the morphology is unique to this breast carcinoma. It usually has associated vascular invasion on presentation.

19) **D**

The question stem gives a description of the features of nodular colloid goitre. No equivocal features or malignant features are described. This is benign – **Thy2**.

20) **E**

The patient is likely to have inherited **Multiple Endocrine Neoplasia 1** (MEN1) which occurs as an abnormality in the MEN1 gene on Chromosome 11q13. NF1/NF2 abnormalities occur in neurofibromatosis, p16INK4A abnormalities occur in malignant melanoma and BRCA1/BRCA2 abnormalities occur in breast cancer. RET abnormalities can occur in MEN2.

21) **C**

Mitral prolapse (floppy mitral valve) is a cause of sudden death in adults not **mitral stenosis**. Bridging of coronary arteries occurs when an epicardial artery is buried in the cardiac muscle for part of its path (to a depth of at least 5mm and a length of 20mm), which can result in a fatal compression of the lumen. Hypertrophic obstructive cardiomyopathy (HOCUM) and ischaemic heart disease can often have a first presentation with sudden death. Wolff-Parkinson-White syndrome can cause fatal arrhythmias.

22) **A**

10% of GISTs occur in the **oesophagus**. The c-kit gene codes for the abnormal tyrosine kinase. Gleevec/imatinib is a drug that inhibits the tyrosine kinase. Interstitial cells of Cajal are believed to be the origin of GISTs. Size (greater than 2 cm) is a poor prognostic factor.

23) **E**

Explanations A-D are correct. **Cyclical breast pain** is usually a sign of benign disease, and is related to the menstrual cycle.

24) D
Inferred consent is not an expression that is used:

- Explicit consent - explicit explanation, explicit agreement.
- Implied consent - patient action implies agreement.
- Specific consent - agreement to limited action.
- General consent - agreement for a wider range of uses.

25) D
Blood group A has an increased risk of stomach cancer. Classically, Japanese and Chinese have an increased risk of stomach cancer – not individuals from South American. Pickled food containing nitrites are positively associated with stomach cancer. As with several other cancers, low socioeconomic status is a risk factor for stomach cancer. There is no relationship between trimethoprim and stomach cancer.

26) C
The GMC Good Medical Practice guidelines require doctors to maintain a familiarity with their *GMC* registration numbers.

27) D
This is the definition of **bronchiectasis**. Emphysema is a similar process occurring in *alveolar air spaces*.

28) B
This is a standard description of an infection by **pneumocystis carinii (jiroveci)**.

29) C
Bilayered oncocytic epithelium exists in a **Warthin's tumour**. The oncocytoma does have oncocytic cells but not in a bilayer arrangement.

30) E
A-D are inhibin positive ovarian tumours. The **Brenner tumour** is CK8, CK18, EMA and CEA positive.

31) B
The infiltrative atypical melanocytes in the reticular dermis, with no connection to the epidermis of the forearm, suggests a **nodular malignant melanoma**. The key features of a dysplastic naevus are not described.

32) E
Ehlers-Danlos syndrome is a risk factor for aortic dissection; however there is no syndrome named **Ehlers-Dunlus**. A-D are recognized risk factors for aortic dissection.

33) A
The **septic mode** of death occurs as the result of an infection. This is a **natural** cause of death and does not require a coroner's autopsy. B represents a possible suicide and

should be referred to the coroner, C represents an operative death and must be referred to the coroner and D should be referred because most mesotheliomas are caused by industrial/occupational exposure. E represents both a post-operative death and a death following pregnancy.

A death should be referred to a coroner if:
The cause of death is unknown.
The death was violent, unnatural or suspicious. This would include suicide, homicide, road traffic accidents, infant deaths or iatrogenic deaths.
The death may be due to an accident. This includes workplace, domestic and road traffic accidents.
The death may be due to self-neglect or neglect by others. This may include death in nursing homes and those sectioned under the Mental Health Act.
The death may be due to an industrial disease related to the deceased's employment.
The death may be due to an abortion.
The death may be related to an operation.
The death may be related to alcohol, poisoning or drug use.
There is a death in custody.

A death certificate can be issued without referral to a coroner if:
- The patient was seen alive by their doctor within the last 2 weeks.
- The doctors are satisfied that the death was due to natural causes.
- The doctors are reasonably sure of the cause.
- None of the above indications for coroner's referral are met.

34) A
Rabies is a Class 3 infectious agent. Escherichia coli is a Class 2 infectious agent. Lassa fever, Variola and Marburg virus are Class 4 agents.

35) A
The question stem describes the histological appearance of **gynaecomastia**. There are no breast lobules present, no fibrosis or fibroelastosis, no macrophages or dystrophic calcification and no evidence of malignancy.

36) D
Arrhythmogenic right ventricular cardiomyopathy is a recognized cause of sudden death of cardiac aetiology. Similarly mitral prolapse (not mitral stenosis) and aortic stenosis (not aortic regurgitation) are causes of sudden death. Atrial myxoma can cause sudden death by releasing pulmonary emboli. There is no entity termed arterial myxoma.

37) B
The malignant glandular cells are located in the periphery of the lung and are creeping along the alveolar walls in a monolayer. This is **bronchioloalveolar carcinoma**. Atypical adenomatous hyperplasia is pre-malignant. Squamous cell carcinoma is not an adenocarcinoma. Small cell carcinoma and renal cell carcinoma do not typically have an alveolar growth pattern.

38) B

The question stem describes a **synovial sarcoma**. The herring bone architecture and CD99 positivity are useful diagnostic features. Atypical fibroxanthoma would appear to be very pleomorphic. Carcinosarcoma is biphasic. Fibrosarcomas are not usually CD99 positive. Leiomyosarcoma is often SMA positive not CD99 positive.

39) B
In Whipple's disease the jejunal villi are packed with **macrophages**.

40) C
The true vocal cords are lined by **stratified squamous epithelium**. The false vocal cords are lined by ciliated columnar epithelium.

41) B
The weight range of a normal placenta is **350-700g**.

42) E
In cirrhosis, the architecture of the liver is altered to **regenerating nodules** circumscribed by fibrosis. The mechanism includes:
A) Laying down of fine fibrotic fibres (collagen) through the parenchyma that connect the central veins to the portal tracts together with *new vascular* channels.
B) Laying down of collagen in the space of Disse prevents diffusion from sinusoids to the hepatic parencyma (fenestrations obstructed).
C) Changing of perisinusoidal cells (Ito cells) into myofibroblasts that contract and increase vascular resistance.

43) C
Temporal arteritis (giant cell arteritis) is a large vessel arteritis that causes pain in the temple of older individuals; without rapid treatment the effect on the ophthalmic artery can lead to blindness.

44) C
Alpha-synuclein is deposited in Lewy body dementia. Tau protein and A-beta protein are deposited in Alzheimer's disease. Amyloid is deposited in a range of diseases including Alzheimer's disease. Alphabeta-crystallin is expressed in renal carcinoma.

45) A
Bronchopulmonary carcinoid prognosis is worsened by lymphatic invasion, increased mitotic count, necrosis and larger size. **Cell density** is not a prognostic factor.

46) B
This is the description of Langerhan's cell histiocytosis. Under electron microscopy the proliferating cells demonstrate **Birkbeck granules**. The proliferating cells are CD1a positive not CD2a positive. Similarly the cells are S100 positive not HMB45 or NK1C3. The Langerhans cells are CD15 negative.

47) A
Maternal smoking is the largest risk factor for premature deliveries.

48) D

CPD does require the collection, completion and submission of documents. Even in a well organized department this may require **some more work**.

The following is a description from the Royal College of Pathology website:

CPD has three main components, namely the knowledge, skills and attributes that a doctor needs for professional practice. As individuals, we are responsible for identifying our learning needs and those of the organisation for which we work through the appraisal process.

There is good evidence that effective CPD can:

  * lead to improved patient care
  * keep us up to date with rapidly expanding knowledge base
  * provide evidence for annual appraisal and revalidation
  * enhance personal development
  * prepare us for new roles, e.g. managerial
  * help us comply with employers' and professional bodies' requirements
  * contribute to clinical governance
  * maintain clinical effectiveness at a high level

49) C

Glutaraldehyde is the fixative used on a renal biopsy specimen being prepared for electron microscopy. Bouin's solution is used for a wide range of purposes, including the preparation of prostatic or testicular specimens.

50) E

Acetylcholinesterase staining is used to identify the hypertrophic extrinsic nerve fibres that can assist pathologists in the diagnosis of Hirschprung's disease.

51) I

The rosettes with fibrillary cores, calcification, necrosis and ganglion-like bodies are all suggestive of **neuroblastoma**.

52) A

Desmoplastic small round cell tumour is biphasic. The immunohistochemical panel is diagnostic for this tumour.

53) C

Lymphoblastic lymphoma is composed of discohesive cells with scanty cytoplasm. This T cell lymphoma is usually LCA, CD7 and TdT positive. It is occasionally also positive for CD99.

54) D

The t(11;22)(q24;q12) translocation occurs most commonly in **Ewings sarcoma** and peripheral neuroectodermal tumours, but can also occur in neuroblastomas. The clinical scenario, including tumour site, and histological description are most consistent with Ewing's sarcoma.

55) B

MyoD1 and MSA are skeletal muscle markers and are positive in **rhabdomyosarcoma**.

56) H
Aortic stenosis and **mitral valve regurgitation** (mitral prolapse) are recognised causes of sudden death in an adult. Marfan's syndrome predisposes to mitral valve regurgitation.

57) E
Aortic stenosis (aortic valve stenosis) classically produces an ejection systolic murmur. A significant component in the pathogenesis of aortic stenosis is calcification of the valve; this results in an increase the prevalence of aortic valve stenosis with advancing age.

58) F
Idiopathic aortic root dilation (annular ring dilation) is a common cause of aortic regurgitation. The dilation of the root means that the valve leaflets are not large enough to meet at the centre of the lumen; the valve cannot completely close and so regurgitation occurs.

59) C
The **tricuspid valve** is a common site of infective endocarditis. Venous borne infective agents will reach this heart valve *first*.

60) K
Fallot's tetraology consists of right ventricular hypertrophy, ventricular septal defect, overriding aorta and pulmonary stenosis. The overriding aorta and hypertrophy of the cardiac muscle (infundibular stenosis) directly contribute to **pulmonary valve stenosis**.

61) L
This is the panel that would generally identify a primary **renal cell carcinoma**.

62) H
This is the panel that would generally identify a primary **lung adenocarcinoma**.

63) A
This is the panel that would generally identify a primary **pancreatic adenocarcinoma**.

64) B
This is the panel that would generally identify a **bladder (transitional cell) carcinoma**.

65) F
This is the panel that would generally identify a primary **ovarian adenocarcinoma**.

66) A
The arteriographic centrifugal filling pattern and central stellate scar are strongly suggestive of **focal nodular hyperplasia**.

67) K
Hepatitis A has a 15-40 day incubation period. Hepatitis B has a 50-180 day incubation period and Hepatitis C has a 40-55 day incubation period.

68) D
Mallory bodies and zone 3 liver cell ballooning are features of steatohepatitis with an alcohol aetiology (**alcoholic steatohepatitis**).

69) J
The Shikata-Orcein stain is positive in the presence of copper salts. Copper deposition occurs throughout the body in **Wilson's disease**; cirrhosis of the liver is a common consequence of longstanding disease.

70) G
Primary sclerosing cholangitis is associated with ulcerative colitis. Onion skinning occurs as a result of concentric fibrosis.

71) L
The question stem describes koilocytes, binucleate cells and keratotic cells. These are features of **HPV infection**.

72) E
Clusters of syncytial squamous cells containing neutrophil microabscesses are a **reparative** feature.

73) H
Markedly dysplastic squamous cells and a tumour diathesis are described. These features are likely to indicate the presence of a **squamous cell carcinoma**.

74) J
The question describes rosettes of endocervical cells; these rosettes are consistent with **CGIN**.

75) K
There are background candidal hyphae. However the point of diagnostic concern is the scalloped clusters of atypical cells containing mucin, which also show nuclear atypia. In a woman of this age one should be concerned about an **endometrial carcinoma**.

76) A
In 85% of cases of Ewing's sarcoma there is a **t(11;22)(q24;q12)** translocation. The result is chimeric protein that constitutively stimulates cell proliferation.

77) B
Mutation in the **mHTT** protein (Huntingtin protein) causes Huntington's disease.

78) F
50% of follicular thyroid carcinomas are caused by RAS oncogenes. **NRAS** is the most common of these mutations.

79) K

85% of the cases of Adult Polycystic Kidney Disease are due to **PKD1** gene mutations on chromosome 16; 16p13.3. Almost all of the remaining 15% of cases are due to mutations on chromosome 4, at the PKD2 gene.

80) D

Goodpasture's syndrome is believed to be the result of a type II hypersensitivity reaction against alpha-3 chain of type IV collagen in the glomerular basement membrane. Hence **Anti-GBM COL4-A3** antigen is key in the pathogenesis of the disease.

81) I

Macrophages have a key role in granulomatous disease and produce a range of inflammatory mediators, which control and coordinate chronic inflammation.

82) E

E-selectin is expressed on endothelium and allows leukocyte binding prior to transmigration.

83) A

PECAM-1 (CD31) belongs to the immunoglobulin family of proteins and facilitates transmigration through vascular walls.

84) N

Nitric oxide is synthesized by the enzyme nitric oxide synthase, which is present in endothelial, neuronal cells and macrophages. **Nitric oxide** acts on vascular smooth muscle to cause vasodilation.

85) F

Chemokines attract leukocytes along their concentration gradient by acting as chemotactic stimuli.

86) M

The two commonest causes of cirrhosis are viral hepatitis and **alcoholic liver disease**; the latter is more common in the West.

87) B

Types I, III and IV collagen are represented in the normal or cirrhotic human liver. **Type II collagen** is not present in the human liver.

88) J

Primary haemochromatosis causes iron deposition in the skin and pancreas, leading to diabetes mellitus and the characteristic skin pigmentation.

89) H

Hepatitis A is a self-limiting disease caused by a viral infection; it is an acute disease and does not lead to cirrhosis.

90) E

The question stem describes the characteristic appearances of Kayser-Fleischer rings that are pathognomonic for **Wilson's disease**.

91) **D**
The **extrinsic** coagulation pathway is triggered by tissue damage. However, the extrinsic and intrinsic pathways are interconnected, for example, a tissue factor-factor VIIa complex can activate factor IX in the intrinsic pathway.

92) **E**
Phospholipid complexes and calcium ions act as cofactors that facilitate multiple steps in the coagulation cascades.

93) **H**
Haemophilia A is an X-linked recessive disease that is caused by a reduction in the amount or activity of **factor VIII**.

94) **A**
Plasmin is a proteolytic enzyme that acts on **fibrin** to cause fibrinolysis.

95) **I**
Christmas disease (Haemophilia B) occurs as a result of an aberrant factor **IX** gene that is located on the X chromosome (Xq27.1-q27.2) and is inherited as an X-linked recessive trait.

96) **E**
Ankylosing spondylitis is a seronegative arthropathy that results in ossification of vertebral and sacroiliac joints. The characteristic radiological appearance is called "bamboo spine."

97) **J**
Systemic lupus erythematosus is frequently associated with a lupus nephritis that causes glomerular changes. "Full house" immunological deposition includes C3, C1q, IgA, IgG and IgM.

98) **H**
Kawasaki disease is a medium vessel vasculitis that can affect the coronary arteries in children. Consequently, coronary artery aneuryms are not uncommon.

99) **L**
The question stem gives a description of **temporal arteritis** (giant cell arteritis, cranial arteritis). This is a large vessel arteritis that is treated as a medical emergency; treatment is usually initiated before receiving the results of blood tests.

100) **A**
Churg-Strauss syndrome is a small vessel vasculitis that causes asthma and eosinophilia.

101) **H**
Aplastic anaemia causes pancytopaenia. It has a range of causes including organic compounds such as benzene.

102) **A**

Haemoglobin is composed of two alpha and two beta subunits.

103) **G**

The autoimmune attack on intrinsic factor prevents the binding and absorption of B12 to cause a macrocytic anaemia; this is **pernicious anaemia**.

104) **F**

Sickle cell anaemia is a known cause of nephrotic syndrome. Sickle cell anaemia is commoner in individuals of Afro-caribbean ethnicity.

105) **B**

Myoglobin is a monomeric oxygen binding protein in skeletal muscle.

106) **J**

The large fibrotic anthracotic nodules described in the question stem are characteristic of **progressive massive fibrosis**, a complicated form of coal worker's pneumoniocosis.

107) **A**

Acute respiratory distress syndrome can be triggered by numerous stimuli including infective causes. The respective histological appearance is of diffuse alveolar damage; this is characterized by hyaline membranes.

108) **E**

Caplan syndrome occurs in individuals with rheumatoid disease (rheumatoid arthritis) who respond to coal dust by undergoing interstitial fibrosis of the lungs and experiencing shortness of breath.

109) **L**

Farmer's lung is a type III hypersensitivity reaction to thermophilic actinomyces that proliferate in warm hay.

110) **G**

The question stem implies that the cause of the lung pathology are the amphibole fibres; these are the asbestos fibres that penetrate most deeply into the lung and trigger the fibrosis. This is the pathogenesis of **asbestosis**.

111) **A**

The intractable peptic ulcers are consistent with Zollinger-Ellison syndrome, which is frequently caused by a pancreatic gastrinoma. The hypercalcaemia and cervical triangle mass are consistent with parathyroid adenoma/hyperplasia. The anterior pituitary mass is consistent with an anterior pituitary adenoma. The simplest explanation for all of these signs and symptoms is the presence of Multiple Endocrine Neoplasia type 1 (**MEN1**) in this patient.

112) **F**

The question stem describes features of carcinoid syndrome; the lung tumour is a **carcinoid tumour** that secretes 5-HT and ultimately causes the flushing. The core features of carcinoid syndrome are flushing, diarrhoea, wheeze and abdominal pain. Consequently, it is frequently referred to as the *toilet* tumour.

113) I
Gastrinomas secrete gastrin and cause the peptic ulcers of Zollinger-Ellison syndrome.

114) B
The most likely diagnosis is Multiple Endocrine Neoplasia type 2A (**MEN2A**). This disease is characterized by medullary thyroid carcinoma, phaechromocytoma and either parathyroid adenoma or hyperplasia.

115) C
MEN2B is characterised by mucosal neuromas as well as medullary thyroid carcinoma and phaeochromocytoma. The simplest explanation for the information in the question stem is the inheritance of MEN2B.

116) J
The question stem describes a scenario consistent with a chronic **subdural haematoma** that classically occurs in the elderly after head trauma.

117) I
The question stem describes an **epidural haemorrhage**.

118) D
Meningiomas can occur as late complications of head trauma.

119) B
The question stem describes a **contrecoup** injury. It represents secondary trauma between the brain and the posterior cranium.

120) F
Obstruction of the flow of cerebrospinal fluid through the foramen of Luschka can cause **hydrocephalus**.

121) A
Fibroadenomas are the most common breast tumours in adolescent women. They are well defined, mobile and usually less than 4cm in maximum extent.

122) F
The location and components of the lesion described in the question stem are consistent with a **nipple adenoma**. Nipple adenomas usually show an adenosis pattern, sclerosis with pseudoinvasive features, sclerosing papillomatosis and infiltrative epitheliosis. The lesion may extend to or erode through the epidermis. Clinically, nipple adenomas present with bloody or serous nipple discharges.

123) J
The ductular atypia with comedo necrosis and calcification, without an invasive focus, is most consistent with high grade **ductal carcinoma in situ**.

124) **B**

A high grade **Phyllodes tumour** has sarcomatous characteristics because of the malignant behaviour of the mesenchymal component.

125) **I**

The question stem describes the classical form of **medullary carcinoma** of the breast. The syncytial cells are large pleomorphic cells. The carcinoma has a good prognosis.

PAPER 3

Part 1 EXAMINATION – Paper 3

1) Which of the following statements is not true in relation to Barrett's oesophagus?

(A) It represents glandular metaplasia of the lower oesophageal mucosa.
(B) It is due to gastro-oesophageal reflux disease.
(C) It is responsible for a marked rise in the incidence of oesophageal
   adenocarcinoma over the last three decades.
✓(D) It is commoner in individuals with Helicobacter pylori infection of the stomach.
(E) It is commoner in obese individuals.

2) Short segment columnar lined oesophagus (CLO) or Barrett's oesophagus refers to glandular metaplasia in the lower oesophagus less than the following lengths:

✓(A) 3 cm
(B) 4 cm
(C) 5 cm
(D) 6 cm
(E) 12 cm

3) The recommended classification of CLO in Western countries includes all of the following categories with the exception of:

(A) Negative for dysplasia
(B) Indefinite for dysplasia
(C) Low grade dysplasia
✓(D) Moderate grade dysplasia
(E) High grade dysplasia

4) A 25-year-old man presents at hospital demonstrating bilateral cervical lymphadenopathy with a three month history and night sweats with a six month history. An excision biopsy of an enlarged lymph node shows capsular and interstitial fibrosis. The cellular population is abnormal, including scattered large binucleate cells with prominent nucleoli consistent with Reed Sternberg cells. Immunohistochemistry is likely to demonstrate which of the following antigens?

(A) CD45
✓(B) CD30
(C) CD20
(D) CD5
(E) EMA

5. The following statements are true in relation to neuroblastoma, except:

(A) It is the most common malignant tumour of childhood.
✓(B) It rarely occurs in females.
(C) Histologically, Homer-Wright pseudo-rosettes may be seen.
(D) Overall genomic pattern, as assessed by array-based karyotyping, is prognostically important.
(E) Urine catecholamine levels may be elevated.

6) The following statements are true in relation to the clinical behaviour of Wilms tumour (nephroblastoma), except:

(A) It rarely occurs in adults.
(B) The majority occur in otherwise normal children.
(C) The prognosis is excellent, with a five year survival rate of approximately 90%.
(D) Only 5% of cases are bilateral.
(E) Metastases are typically to the liver.

7) The following statements are true in relation to the pathology of Wilms tumour (nephroblastoma), except:

(A) There is a triphasic pattern including metanephric blastema, stromal and epithelial derivatives.
(B) The stroma may contain striated muscle, cartilage, bone or fat.   *Tendono*
(C) Wilms tumour shows two broad prognostic categories based on pathological appearances.
(D) Loss of heterozygosity (LOH) for chromosomes 1p and 16q identifies a subset of Wilms tumour patients who have a significantly increased risk of relapse and death.
(E) Mutations of the WT1 gene on chromosome 11 are observed in all Wilms tumours.

8) The following immunohistochemical stain is a sensitive and specific marker for precursor acute lymphoblastic leukaemia/lymphoma:

(A) Terminal deoxynucleotidyl transferase (TdT)
(B) CD45
(C) CD79a
(D) CD15
(E) Myeloperoxidase

9) The following statements are true of matrix metalloproteinases, except:

(A) They are frequently up regulated in human tumours.
(B) They are inhibited by molecules called tissue inhibitors of metalloproteinases.
(C) They are produced not by tumour cells themselves but by associated stromal cells.
(D) As well as being involved in extracellular matrix degradation, MMPs are involved in tumour growth regulation, apoptosis and angiogenesis.
(E) They may sometimes be expressed in normal tissues.

10) In advanced osteoarthritis, a femoral head removed for prosthetic replacement is likely to show the following changes, except:

(A) Eburnation
(B) Osteophyte formation at the margins of the femoral head.
(C) Formation of subchondral bone cysts.
(D) A dense lymphocytic and plasma cell infiltrate within the synovium.
(E) Fissure formation within the articular cartilage.

11) A 10-year-old boy presents with a 2 cm nodule that has grown rapidly over a week. Histologically, there is a zonal pattern, with a central myxoid area containing myofibroblasts, surrounded by a more fibrous area containing spindle cells. The myofibroblasts express muscle specific actin, smooth muscle actin and desmin, but are negative for myogenin as well as CD34 and cytokeratins. The most likely diagnosis is:

(A) Nodular fasciitis
(B) Synovial sarcoma
(C) Infantile myofibromatosis
(D) Proliferative fasciitis
(E) Low-grade fibromyxoid sarcoma

12) Blood tests may detect the presence of placental alkaline phosphatase (PLAP) in fifty percent of cases of which malignant neoplasm?

(A) Seminoma
(B) Choriocarcinoma
(C) Yolk sac tumour
(D) Embryonal carcinoma
(E) Teratocarcinoma

13) Risk factors for the development of a testicular germ cell neoplasm include all of the following except:

(A) Caucasian origin
(B) Previous mumps orchitis
(C) Cryptorchidism
(D) Age 15 to 40 years
(E) Down's syndrome

14) Meigs' syndrome refers to the combination of:

(A) Ascites, pleural effusion and ovarian tumour.
(B) Dystonia and blepharospasm.
(C) Gonadal dysgenesis, nephropathy, Wilms tumour.
(D) Multiple colonic polyps, thyroid cancer and epidermal cysts.
(E) Sudden onset of fever, leukocytosis, and tender, erythematous, well-demarcated papules and plaques.

15) Alpha-fetoprotein may be raised in the following tumours, with the exception of:

(A) Hepatocellular carcinoma
(B) Yolk sac tumour
(C) Neuroblastoma
(D) Hepatoblastoma
(E) Embryonal carcinoma

16. The following statements about hereditary non-polyposis colorectal carcinoma are correct, except one:

(A) Tumours typically affect the right side of the colon.
(B) They most frequently show mucinous differentiation
(C) It is also referred to as Lynch syndrome
(D) Inheritance is through an autosomal recessive pattern
(E) It is sometimes also associated with a high risk of endometrial, ovarian and gastric carcinomas

17) Which of the following small round blue cell tumours of childhood is common in patients with familial adenomatous polyposis (FAP)?

(A) Retinoblastoma
(B) Hepatoblastoma
(C) Ewing's sarcoma
(D) Wilms tumour
(E) Neuroblastoma

18) Sarcoma botryoides is a manifestation of which malignancy?

(A) Embryonal rhabdomyosarcoma
(B) Alveolar rhabdomyosarcoma
(C) Primitive neuroectodermal tumour
(D) Ewing's sarcoma
(E) Desmoplastic small round cell tumour
(F) Epithelioid sarcoma

19) Which of the following immunohistochemical stains is pathognomonic, in the correct context, for gastrointestinal stromal tumour (GIST), and can be used to guide therapy?

(A) CD34
(B) CD43
(C) CD117
(D) cyclin D1
(E) CD30 (ki-1)

20) Alport syndrome may cause the following type of glomerulonephritis:

(A) Membranous glomerulonephritis
(B) Focal segmental glomerulosclerosis
(C) IgA nephropathy
(D) Membranoproliferative/mesangiocapillary glomerulonephritis
(E) Rapidly progressive glomerulonephritis

21) Membranous glomerulonephritis may be caused by all of the following, except:

(A) SLE
(B) Penicillamine therapy
(C) Malaria
(D) Colonic carcinoma
(E) Digoxin therapy

22) The following statements are true in relation to sudden unexpected death in infancy (SUDI), except:

(A) An adequate explanation for death is revealed by a detailed autopsy.
(B) No adequate explanation for death is revealed by a detailed autopsy.
(C) Infants suffering SUDI show a different nasopharyngeal bacterial population to normal infants.
(D) Splenic culture is a standard post-mortem investigational technique in these cases.
(E) Following culture of CSF, pure growth of a bacterial pathogen, associated with increased inflammatory cells, may safely be regarded as the most likely cause of death.

23) Peak incidence of SUDI and SIDS (sudden infant death syndrome) occurs at the following stage of life:

(A) Under one month of age.
(B) 2 to 3 months of age.
(C) 4 to 5 months of age.
(D) 6 to 8 months of age.
(E) 8 to 12 months of age.

24) The most frequent causative organism of septicaemia and meningitis in infants under one month old is:

(A) Group B Streptococcus
(B) Mycobacterium tuberculosis
(C) Vibrio species
(D) Neisseria meningitidis
(E) Haemophilus influenzae

25) Epstein-Barr virus infection may be detected by in situ hybridisation in 30 to 40% of cases of the following lymphoma:

(A) Precursor T-cell acute lymphoblastic leukaemia/lymphoma
(B) Precursor B-cell acute lymphoblastic leukaemia/lymphoma
(C) Endemic Burkitt's lymphoma
(D) HIV-associated Burkitt's lymphoma
(E) NK/T cell lymphoma of nasal type

26) The commonest lymphoma in the paediatric age group is:

(A) Burkitt's lymphoma
(B) Classical Hodgkin's lymphoma
(C) Nodular lymphocyte predominance Hodgkin's lymphoma
(D) Lymphoblastic lymphoma
(E) Peripheral T-cell lymphoma

27) Kasabach-Merritt syndrome refers to a cavernous haemangioma that causes:

(A) High-output cardiac failure

(B) Thrombocytopenia
(C) Rupture with massive haemorrhage
(D) Neutropenia
(E) Malignant transformation to angiosarcoma

28) The following statements are true in relation to anaplastic large cell lymphoma except:

(A) The vast majority of cases are positive for CD30
(B) There is a predominance of cases in males
(C) The disease frequently presents extranodally
(D) A minority of cases are of B-cell lineage
(E) Most cases show a characteristic t(2;5) translocation

29) Which of the following cell types is not seen in classical Hodgkin's lymphoma?

(A) Hodgkin cell
(B) Reed-Sternberg cell
(C) Mummified cell
(D) Popcorn cell
(E) Lacunar cell

30) The following statements are true in relation to metastatic adenocarcinoma of unknown origin, except:

(A) Median survival is under six months.
(B) CA125 detection in serum may be helpful in suggesting a likely primary site.
(C) PSA detection in serum may be helpful in suggesting a likely primary site.
(D) CA19-9 detection in serum may be helpful in suggesting a likely primary site.
(E) Electron microscopy can be valuable in distinguishing possible primary sites for adenocarcinomas of unknown origin.

31) The following malignancy occurs with increased frequency in HIV/AIDS:

(A) Gastric carcinoma
(B) Pancreatic carcinoma
(C) Carcinoma of the tongue
(D) Glioblastoma multiforme
(E) Acute myeloid leukaemia

32) The following statements are correct in relation to cutaneous basal cell carcinoma, except:

(A) 80% of cases occur in the head and neck area.
(B) There is a strong association with Gorlin's syndrome.
(C) 30% of cases occurring in non-sun exposed areas of the skin.
(D) Around 30% of Caucasians develop basal cell carcinoma at some stage in life.
(E) There is a strong association with Gartner's syndrome.

33) The following tumours are important diagnostic criteria of Gorlin's syndrome, except:

(A) More than 2 BCCs or 1 BCC in an individual younger than 20 years.
(B) Odontogenic keratocysts of the jaw.
√(C) Colonic carcinoma in an individual younger than 20 years.
(D) Calcification of the falx cerebri.
(E) Bifid, fused, or splayed ribs.

34) The following lesions may be seen in the heart in rheumatic fever:

(A) Howell-Jolly bodies
(B) Psammoma bodies
√ (C) Aschoff bodies
(D) Negri bodies
(E) Pick bodies

35) A primary lesion caused by mycobacterium bacilli (tuberculosis) developed in the lung of a previously uninfected individual is referred to as a:

√ (A) Ghon focus
(B) Ghon complex
(C) Simon's focus
(D) Koch complex
(E) Pott's focus

36) The following statements are true in respect of mycobacterium tuberculosis, except:

(A) The organism is an obligate aerobe.
(B) About 90% of people infected with Mycobacterium tuberculosis have asymptomatic, latent TB infection.
(C) Lower body mass index is a risk factor.
(D) Diabetes mellitus is a risk factor.
√(E) Negative ZN staining excludes infection.

37) The following statements are true of Kikuchi's disease, except:

(A) Most cases are self-limiting and resolve without treatment.
(B) Most cases occur in young adults.
(C) Lymph node biopsy is required for diagnosis.
(D) Lymph node necrosis is a characteristic feature.
√(E) Serology shows anti-double stranded DNA antibodies.

38) The following molecules are involved in the process of apoptosis, except:

(A) fas ligand (FasL)
(B) tumour necrosis factor (TNF)
(C) p53 protein
√(D) CD103

(E) proteolytic caspases

39) In malignant melanoma, the following are important prognostic indicators, except:

(A) Presence of associated lymphocytic infiltrate
(B) Anatomical site of the melanoma
(C) Diameter of the melanoma
(D) Lymph node involvement
(E) Presence of ulceration

40) The following statements are true in respect of frozen section diagnosis, except:

(A) It is used commonly to identify metastases in sentinel lymph nodes.
(B) It forms an important part of the performance of Moh's surgery.
(C) The specimen is embedded in a gel like medium known as Optimum Cutting Temperature Compound.
(D) Some substances lost during paraffin processing may be demonstrated e.g. lipids, on frozen sections.
(E) In careful hands, the technique may produce equivalent quality results to paraffin processing.

41) The following histochemical stain may be used to demonstrate fat:

(A) Congo red
(B) Sirius red
(C) Oil Red O
(D) Safranin
(E) Methyl green pyronin

42) The following histochemical stain may be used to demonstrate fungal organisms:

(A) Bismarck brown
(B) Crystal violet
(C) Silver stain
(D) Sudan black
(E) Ziehl-Neelsen stain

43) The following term best describes nuclear fragmentation from a histological point of view:

(A) Apoptosis
(B) Karyorrhexis
(C) Pyknosis
(D) Karyolysis
(E) Necrosis

44) The adaptive process of cellular transformation, from one type of epithelium into another due to a physiological or pathological stressor, is termed:

(A) Dysplasia

✓(B) Metaplasia

(C) Hypertrophy

(D) Hyperplasia

(E) Anaplasia

45) The following fixative is used to prepare tissue for electron microscopy:

(A) B5

✓(B) Glutaraldehyde

(C) Bouin's fluid

(D) Zenker's fluid

(E) Neutral buffered formalin

46) The following histochemical stain can be used for the histopathological diagnosis of malaria:

(A) Warthin-Starry stain

✓(B) Giemsa stain

(C) Silver stain

(D) Van Gieson's Stain

(E) Methylene blue

47) The following parasitic organism is associated with an increased incidence of carcinoma of the biliary tract:

✓(A) Clonorchis sinensis

(B) Paragonimus westermani

(C) Ascaris lumbricoides

(D) Dracunculus medinensis

(E) Taenia saginata

48) Löffler's syndrome is associated with infection with the following parasitic organism:

(A) Echinococcus granulosus

(B) Onchocerca volvulus

✓(C) Ascaris lumbricoides

(D) Loa loa

(E) Toxoplasma gondii

49) The following statements are true of toxoplasmosis, except:

(A) Around one third of the world's population carry the infection.

✓(B) Necrotising granulomas are a feature of the lymphadenitis produced by the disease.

(C) The carrier state may manifest as a variety of psychological/psychiatric disorders.

(D) Fatalities are rare except in the immunocompromised and fetuses.

(E) Follicular hyperplasia is a feature of the lymphadenitis produced by the disease.

50) The following is the commonest site of hypertensive brain haemorrhage:

(A) Cerebellum
(B) Putamen
(C) Thalamus
(D) Cerebrum
(E) Pons

Pair each of the descriptions below to the sarcoma that is the best match:

A) Granulocytic sarcoma
B) Alveolar rhabdomyosarcoma
C) Embryonal rhabdomyosarcoma
D) Ewing's sarcoma
E) Malignant rhabdoid tumour

F) Small cell osteosarcoma
G) Renal sarcoma
H) Synovial sarcoma

51) This highly aggressive tumour has a poor prognosis, and tends to occur in children less than two years of age. It was first described as a variant of Wilms tumour of the kidney in 1978. Around 80% of cases show abnormalities of 22q11.

52) This is not a true sarcoma in that it is not of mesenchymal derivation. There is usually immunoreactivity for one or more of CD15, CD68, CD43, CD34 and myeloperoxidase.

53) This tumour, most common in older children and teenagers, shows appearances resembling those of a typical 10-to-12-week embryo, with formation of pseudo-spaces lined by poorly cohesive cells.

54) A malignant round-cell tumour that most commonly occurs in the pelvis, the femur, the humerus, and the ribs. It is caused by a translocation between chromosomes 11 and 22.

55) A histologically variable malignancy. It has a spindle cell variant most common in the head and neck and testis, and a botryoid variant which classically manifests within coelomic cavities.

Match each of the descriptions below with the most appropriate sarcoma:

A) Malignant peripheral nerve sheath tumour

F) Myxofibrosarcoma

B) Synovial sarcoma

G) Chondrosarcoma

C) Inflammatory myofibroblastic tumour

H) Intimal sarcoma

D) Liposarcoma

E) Malignant fibrous histiocytoma

56) This sarcoma occurs most commonly in teenagers, within the paediatric age group. It is often found adjacent to joints; classically the knee joint. Approximately 30% of cases have a biphasic pattern. Positive immunohistochemical staining for cytokeratins is of great use diagnostically.

57) Although common in adults, this tumour may occur in teenagers, where it is frequently associated with an autosomal dominant-inherited disease. It is NSE and S100 positive.

58) Sometimes referred to as a "dustbin diagnosis" including poorly differentiated sarcomas where a definite histogenesis cannot be assigned, or may represent de-dedifferentiation of a number of sarcomas.

59) Most cases of this tumour occur before the age of 20. Numerous lymphocytes and plasma cells are typical histologically. Clinically the behaviour is relatively low-grade, with recurrence in about 25% of cases. It is sometimes called inflammatory pseudotumour.

60) This tumour is most frequent in middle-aged and older adults in deep-seated locations and is the second most common of all soft-tissue sarcomas. The tumour can grow into a large painless mass that is yellow and lobulated. It is often S100 positive.

Match each of the following descriptions below to the one of the salivary gland tumours, A-J, which best fits it:

A) Polymorphous low-grade adenocarcinoma
B) Lymphoepithelial carcinoma
C) Acinic cell carcinoma
D) Epithelial-myoepithelial carcinoma
E) Adenoid cystic carcinoma

F) Warthin tumour
G) Mucoepidermoid carcinoma
H) Pleomorphic adenoma
I) Oncocytoma
J) Squamous cell carcinoma

61) This is typically a slowly enlarging mass in the parotid region. It is a malignant tumour in which at least a proportion of cells exhibit serous acinar cell differentiation and show DPAS positivity.

62) An aggressive basaloid carcinoma with a tendency to spread along nerves. Glandular lumina are present. It shows three major variant histologic growth patterns: cribriform, tubular and solid.

63) A relatively indolent malignancy, most cases arising from minor salivary glands. Although the tumour cells tend to be cytologically bland and monomorphic, it can form a wide variety of architectural patterns.

64) This rare carcinoma shows a biphasic pattern, with glandular structures lined by epithelial cells and surrounded by an outer layer of (clear) myoepithelial cells. Ductular proliferation is common. It has a slight female preponderance and tends to occur in individuals beyond the sixth decade.

65) This carcinoma is rare in the West but has a high incidence in China, where EBV is implicated as the aetiological agent.

Match each of the following descriptions below, 66-70, to the sex cord-gonadal stromal tumour that best describes it:

A) Granulosa cell tumour

B) Thecoma

C) Fibroma

D) Sertoli cell tumour

E) Leydig cell tumour

F) Gynandroblastoma

G) Steroid cell tumour

H) Sex cord tumour with annular tubules

I) Fibrosarcoma

66) This unilateral ovarian tumour tends to present in women in the 50-55 year old age group with post menopausal bleeding per vaginam. The nuclei of the tumour cells each have a longitudinal groove. The tumour cells are usually CD99 positive.

67) This ovarian tumour is almost invariably benign and unilateral. Microscopically, the tumour has a tubular organization and multiple hyaline bodies are present. It typically secretes androgens.

68) This unilateral 10 cm ovarian tumour is firm and white and occurred in a patient with ascites. No mitotic figures were observed on microscopic examination. Such a tumour may be associated with Meig's syndrome.

69) This ovarian malignancy is a densely cellular spindle cell neoplasm with moderate to severe cytological atypia and a high mitotic rate. Atypical mitotic figures, haemorrhage and necrosis are evident.

70) The majority of cases of this tumour are found in men, usually at 5-10 years of age or in middle adulthood (30-60 years). Children typically present with precocious puberty and painless testicular enlargement. The constituent cells are polygonal cells with eosinophilic cytoplasm and distinct cell borders.

Match each of the statements below, 71-75, to the most appropriate central nervous system pathology:

A) Wernicke's encephalopathy
B) Aseptic meningitis
C) Lewy body dementia
D) Huntington's chorea
E) Primary CNS lymphoma

F) Metastatic adenocarcinoma
G) Metastic malignant melanoma
H) Alzheimer's disease
I) Stroke
J) Transient ischaemic attack

71) This is the second commonest cause of cognitive decline, dementia, in elderly people. There is a characteristic change in the pallor of the substantia nigra.

72) This disease is characterized by brief, irregular, involuntary movements that can be misininterpreted as restlessness. There is a subsequent progressive cognitive decline. The disease is inherited in an autosomal dominant manner.

73) This condition is strongly associated with HIV infection, particularly in patients whose CD4 count is low. It has also been increasing rapidly in incidence in immunocompetent people in recent years, for unknown reasons.

74) This condition occurs as a result of thiamine deficiency and is frequently seen in chronic alcoholism. Histologically, haemorrhagic necrosis of the mammillary bodies may be seen acutely whereas cases with more chronic onset exhibit mammillary body gliosis. Ataxia, ophthalmoplegia, confusion and impairment may occur in affected individuals.

75) At autopsy, macroscopic examination of the oedematous brain reveals numerous well-defined solid spherical nodules all of which stain black with Masson Fontana.

Match each of the following immunohistochemical staining patterns with the tumour type, A-J, which is the best fit:

A) Carcinomas in general
B) Adenocarcinomas in general
C) Transitional cell carcinoma
D) Lung and thyroid adenocarcinomas
E) Squamous cell carcinoma

F) Renal cell carcinoma
G) Leiomyoma
H) Small cell carcinoma
I) Pancreatobiliary adenocarcinoma
J) Hepatocellular adenocarcinoma

76) AE1/3 positive

77) CAM 5.2 positive

78) CK5/CK6 positive

79) CK20 positive, CK7 negative

80) TTF1 positive

Match each of the following immunohistochemical staining patterns with the tumour type, A-J, which is the best fit:

A) Breast carcinoma
B) Colonic carcinoma
C) Mesothelioma
D) Thyroid adenocarcinoma
E) Clear cell renal cell carcinoma

F) Pancreatic adenocarcinoma
G) Endometrial adenocarcinoma
H) Thymoma
I) Non-small cell lung cancer
J) Cervical carcinoma

81) GCDFP-15 positive

82) Mesothelin positive

83) CDX2

84) CK20 negative, CK7 negative

85) CK20 positive, CK7 positive

Match each of the following immunohistochemical phenotypic patterns to the most appropriate non-Hodgkin's lymphoma:

A) Marginal zone lymphoma
B) CLL/SLL*
C) Mantle cell lymphoma
D) Follicular lymphoma
E) Precursor B cell ALL
(Acute lymphoblastic leukaemia)

F) Mycosis fungoides
G) T-cell leukaemia
H) Anaplastic large cell lymphoma
I) Burkitt's lymphoma
J) Pleomorphic T cell lymphoma

86) CD20 positive, CD5 positive, CD43 positive, CD23 positive, cyclin D1 negative, CD10 negative.

87) CD20 positive, CD5 negative, CD43 positive, CD23 negative, cyclin D1 negative, CD10 negative.

88) CD20 positive, CD5 positive, CD43 positive, CD23 negative, cyclin D1 positive, CD10 negative.

89) CD20 positive, CD5 negative, CD43 negative, CD23 negative, cyclin D1 negative, CD10 positive.

90) CD20 negative, CD79a positive, CD34 positive, TdT positive, cyclin D1 negative, CD10 positive.

*Chronic lymphocytic leukaemia/Small lymphocytic lymphoma

Match each of the following inborn errors of metabolism with the statement that best fits it:

A) An autosomal recessive genetic disorder in which copper accumulates in brain and liver, manifesting as neurological and psychiatric symptoms and liver disease.
B) An X-linked recessive disorder of copper metabolism, presenting usually in infancy with sparse and coarse hair, growth failure, and neurological symptoms.
C) A rare autosomal recessive disorder of bilirubin metabolism producing unconjugated hyperbilirubinemia. Two distinct forms have been described, type 1 and type 2.
D) A rare inherited disorder characterized by a reduction in serum high density lipoproteins (HDL), premature atherosclerosis and enlarged orange coloured tonsils.
E) Hereditary deficiency of the steroid sulfatase (STS) enzyme, which manifests with dry, scaly skin.
F) A rare idiopathic syndrome associated with conjugated hyperbilirubinaemia, usually presenting in infancy or childhood. The liver cells are not pigmented.
G) A rare autosomal recessive lysosomal storage disease, associated with accumulation of cholesteryl esters and triglycerides, usually fatal at a very young age.

91) Tangier disease

92) Rotor syndrome

93) X-linked ichthyosis (XLI)

94) Wolman disease

95) Menkes disease

114

Match each of the following neurological disorders to the statement below that fits it best:

A) Amyloid plaques and neurofibrillary tangles are clearly visible by microscopy in brains of sufferers. This disease is the commonest cause of dementia.

B) The earliest symptoms of this disease are a general lack of coordination and an unsteady gait. This disease is hereditary.

C) Symptoms usually present between the ages of 50-70, and include progressive weakness, muscle wasting, and fasciculations, spasticity or stiffness, and overactive tendon reflexes. Long established disease can result in generalized paralysis, loss of speech and difficulty swallowing.

D) This disease resembles Alzheimer's, but personality change occurs prior to any form of memory loss in most cases. 3R tau protein is the dominant tau protein present and is found in the pathognomonic body.

E) The primary pathology is insufficient stimulation of the motor cortex by the basal ganglia due to reduced dopaminergic activity – this leads to macroscopically observable changes in the substantia nigra.

F) This progressive disease can cause behaviourial changes, frank psychosis and tremor. Kayser-Fleischer rings may be present in the limbus of the cornea.

G) This progressive disease can cause the patient to show a mixed picture of paretic disease and tabes dorsalis.

H) This disease occurs acutely in young patients and can cause alterations in mood, memory and behaviour. Acyclovir is an effective treatment.

96) Parkinson's disease

97) Alzheimer's disease

98) Pick's disease

99) Huntingdon's chorea

100) Motor neurone disease

Match each of the following pathological conditions with the substance that accumulates in the disease:

A) Cholesterol esters
B) Glucocerebroside
C) Sphingomyelin
D) Mallory's hyaline
E) Pseudomelanin

F) G_{M2} ganglioside
G) Sulfatide
H) Alpha – 1-antitrypsin
I) Iron
J) Glycogen

101) Melanosis coli

102) Xanthelasmata

103) Pompe disease

104) Niemann-Pick's disease

105) Alcoholic liver disease

116

Match each of the following neurological disorders to the statement that describes it best:

A) 60% of cases are associated with small cell lung cancer. The disease presents with proximal muscle weakness and autonomic dysfunction that tends to spare the respiratory muscles and facial muscles. Anticholinesterases are ineffective.

B) A spirochaete-caused disorder that may cause various chronic neurological symptoms.

C) The disease is characterised by fluctuating muscle weakness, due to circulating antibodies which block acetylcholine receptors.

D) This is the commonest cause of inherited learning disability. This disorder demonstrates characteristic elongated facial features, prominent ears, hypotonia, speech abnormalities and atypical social development.

E) An autoimmune demyelinating disorder affecting the peripheral nervous system, usually triggered by a viral infection.

F) A demyelinating disease that has unusual neurological symptoms without a clear anatomical focus. Optic neuritis is a characteristic presentation of this disease.

G) 80% of individuals who have this metabolic disease develop peripheral neuropathy after 15 years. The pattern may be sensorimotor neuropathy, autonomic neuropathy and/or multifocal asymmetric neuropathy.

H) This muscle wasting disease is inherited as an X-linked disease and has an underlying abnormality in the DMD gene.

106) Lambert-Eaton myasthenic syndrome

107) Myasthenia gravis

108) Guillain-Barré syndrome

109) Fragile X syndrome

110) Multiple sclerosis

Match each of the following lung tumours with the statement which best describes it:

A) Although variable in location, most tumours are peripherally located. This is an undifferentiated carcinoma lacking distinctive histological features. The tumour usually consists of sheets of nests of large polygonal cells with vesicular nuclei, prominent nucleoli and a moderate amount of cytoplasm. It is a diagnosis of exclusion.

B) Usually peripherally located, this carcinoma produces a pneumonia-like consolidation rather than a more distinct focal mass and shows "lepidic" spread along alveolar walls. Histologically, focal invasion is rarely found.

C) The commonest location of this carcinoma is peripheral. This tumour shows glandular differentiation and approximately 80% of cases demonstrate mucin. Histologically, focal invasion is invariably found.

D) This neuroendocrine tumour may be centrally or peripherally located and is formed of "packets" of uniform polygonal cells with round nuclei, finely granular chromatin, inconspicuous nucleoli and scant to moderate eosinophilic cytoplasm.

E) Usually centrally located, this tumour is comprised of small cells frequently showing nuclear molding and a granular chromatin pattern. Crush artefact is common. Necrosis and mitotic figures are frequent.

F) This non-small cell carcinoma is classically located adjacent to bronchi. Microscopically, intercellular bridges and/or keratinization may be seen.

G) Radiologically this carcinoma shows large multiple well defined spherical masses in the lung fields termed cannonball metastases. It tends to occur in middle aged men.

111) Bronchioloalveolar carcinoma

112) Carcinoid tumour

113) Small cell carcinoma

114) Adenocarcinoma

115) Large cell carcinoma

Match each of the following pathological conditions with the substance that accumulates in the disease:

A) Collagen
B) Homogentisic acid
C) Ferruginous bodies
D) Immunoglobulin
E) Fibrin

F) Thrombin
G) Amyloid
H) Elastin
I) Lipid
J) Copper

116) Asbestosis

117) Ochronosis (dark pigmentation of auricular cartilage)

118) Keloid

119) Mott cells of multiple myeloma

120) Hyaline membranes

Match each of the following pathological entities below with the statement which best describes it:

A) Leiomyoma   F) Eccrine poroma
B) Krukenberg tumour   G) Michaelis Gutman body
C) Peutz-Jegher's tumour   H) Linitis plastica
D) Cylindroma   I) Hyperplastic polyp
E) Masson's tumour   J) Pulmonary hamartoma

121) A well circumscribed lesion that may present as intrabronchial polypoid mass causing obstruction. Microscopy can reveal representative tissue from mesoderm, ectoderm and endoderm. No invasion is noted.

122) This condition occurs as the result of the metastasis of gastric carcinoma cells, with signet ring morphology, to the ovary.

123) Intravascular papillary endothelial hyperplasia is a benign vascular lesion.

124) This is a well defined tumour of the skin which consists of a proliferation of uniform basaloid cells that are PAS positive. Ducts are evident. The proliferating cells are organized into a "moat and hillock" architecture.

125) This is caused by a poorly differentiated diffuse adenocarcinoma infiltrating the wall of the stomach. Gastric, breast and lung metastases to the stomach can have this effect.

ANSWERS

1) D	43) B	85) F
2) A	44) B	86) B
3) D	45) B	87) A
4) B	46) B	88) C
5) B	47) A	89) D
6) E	48) C	90) E
7) E	49) B	91) D
8) A	50) B	92) F
9) C	51) E	93) E
10) D	52) A	94) G
11) A	53) B	95) B
12) A	54) D	96) E
13) E	55) C	97) A
14) A	56) B	98) D
15) E	57) A	99) B
16) E	58) E	100) C
17) B	59) C	101) E
18) A	60) D	102) A
19) C	61) C	103) J
20) B	62) E	104) C
21) E	63) A	105) D
22) A	64) D	106) A
23) B	65) B	107) C
24) A	66) A	108) E
25) D	67) D	109) D
26) B	68) C	110) F
27) B	69) I	111) B
28) D	70) E	112) D
29) D	71) C	113) E
30) D	72) D	114) C
31) C	73) E	115) A
32) E	74) A	116) C
33) C	75) G	117) B
34) C	76) A	118) A
35) A	77) B	119) D
36) E	78) E	120) E
37) E	79) C	121) J
38) D	80) D	122) B
39) C	81) A	123) E
40) E	82) C	124) F
41) C	83) B	125) H
42) C	84) E	

Paper 3 - Answers

1) D

Barrett's oesophagus (columnar lined oesophagus or CLO) is indeed defined as glandular mucosal metaplasia of the lower oesophagus. It is due to gastro-oesophageal reflux disease and is responsible for what has been termed an epidemic of oesophageal adenocarcinoma in the last decade in the Western world. Individuals without **Helicobacter pylori** infection have a higher stomach acid level and are therefore more prone to developing Barrett's oesophagus should they be predisposed to reflux. There is also a much higher incidence of reflux in obese than normal-weight individuals.

2) A

Short segment CLO refers to an area of glandular metaplasia in the lower oesophagus of under **3 cm** in length. This includes projections of cancer epithelium extending upwards from the squamo-columnar junction. It is far commoner than classical CLO, affecting 8 to 17% of the endoscopic population compared with 1 to 2% for the latter. Risk of malignant transformation in Barrett's oesophagus is thought to be approximately proportional to the length of an underlying CLO segment.

3) D

Although dysplasia in CLO was previously classified as mild, **moderate** or severe, this is now obsolete, having been replaced with the current two tier system which shows much better interobserver agreement. Other categories not mentioned in the question text are intra-mucosal carcinoma and invasive adenocarcinoma.

4) B

90 to 95% of cases of Hodgkin's lymphoma show positivity for **CD30** in Hodgkin's and Reed Sternberg cells. The vast majority of cases are negative for CD45 and EMA This negativity is useful in distinguishing classical Hodgkin's lymphoma from nodule lymphocyte predominance Hodgkin's lymphoma. In the latter there is typically expression of these latter two markers by the "lymphocytic and histiocytic" (L+H) or "popcorn" cells. CD20 is positive in around 10-20% of cases of classical Hodgkin's lymphoma. This is important to report since these patients can be treated with Rituximab should there be a need to progress to salvage therapy.

5) B

Amongst malignancies of childhood, only **desmoplastic small round cell tumour rarely occurs in females**. Neuroblastoma is the commonest extracranial solid tumour of childhood and the commonest malignancy among infants. A classical small round blue cell tumour, pseudo-rosettes may be seen. The genomic profile is important from a prognostic point of view. Study of a large series elucidated the following points: Tumours presenting exclusively with whole chromosome copy number changes were associated with excellent survival. Tumours presenting with any kind of segmental chromosome copy number changes were associated with a high risk of relapse.
Within tumours showing segmental alterations, additional independent predictors of decreased overall survival were N-myc amplification, 1p and 11q deletions, and 1q gain. Urinary catecholamine levels are frequently elevated in neuroblastoma, even in preclinical disease.

6) E

Metastases from Wilms tumour **are typically to the lungs rather than the liver**. The other statements are correct: cases are very rare after childhood; 75% of all cases occur in otherwise normal children while 25% are associated with other developmental abnormalities. Wilms tumour is very responsive to therapy, with a five year survival rate of at least 90%. Around 95% of cases are unilateral.

7) E

Mutations of the **WT1** gene on chromosome 11 are observed in around **20% of Wilms tumours**. Wilms tumour does indeed show a triphasic pattern classically, and all the specialised tissues mentioned may be present. Wilms tumour may be separated into two broad prognostic groups based on histological findings:
* Favourable - Contains well developed components mentioned above
* Anaplastic - Shows areas of diffuse anaplasia.
LOH for chromosomes 1p and 16q identifies a group of Wilms tumour patients who have a particularly poor prognosis.

8) A

Terminal deoxynucleotidyl transferase (TdT) is positive in around 95-97% of cases of ALL. Except in rare exceptions it is negative in all other malignancies. CD45, although positive in the vast majority of lymphomas, which derive from mature rather than precursor lymphocytes, is typically negative in ALL. CD79a is normally positive in B cell acute lymphoblastic leukaemia/lymphoma, but is negative in T-cell cases, so cannot be regarded as a sensitive marker for the disease as a whole. CD15 and myeloperoxidase are myeloid markers and are negative in ALL with the exception of unusual biphenotypic cases. CD34 is frequently positive in ALL, but also in acute myeloid leukaemia, so is not a specific marker.

9) C

Matrix metalloproteinases may be produced both **by tumour cells** and associated stromal cells. As well as being able to degrade most components of the extracellular matrix, they are also intimately involved in the regulation of other aspects of tumour growth and metastasis including those mentioned above. They are up regulated in many cancers and are inhibited by tissue inhibitors of metalloproteinases (TIMPs), of which there are four. Although MMPs are typically up regulated in inflammation and neoplasia, some may be expressed in normal tissues.

10) D

Eburnation refers to thickening of the subchondral bone where the cartilage has been denuded. Osteophytes are frequent at the margins of the joint. Subchondral bone cysts or geodes may be produced in advanced osteoarthritis due to synovial fluid entering the bone marrow. **Dense chronic inflammatory infiltrates are not a feature of osteoarthritis**, being rather seen in rheumatoid arthritis. Damage to superficial articular cartilage occurs secondary to loss of proteoglycans as part of osteoarthritis.

11) A

The clinical and histological findings here are characteristic of **nodular fasciitis**. Helpful features include the short duration of onset, the small size, the presence of a zonal pattern and an absence of atypical mitotic figures. Synovial sarcoma does not show a zonal pattern, may have an epithelioid element, and expresses cytokeratins.

Infantile myofibromatosis shows at least focal positivity for myogenin. Proliferative fasciitis is rare in children and is distinguished by the presence of large basophilic ganglion like cells. Low-grade fibromyxoid sarcoma is also rare in this age group. It is composed of monomorphic small spindle cells which form a mixture of fibrous and myxoid tissues, and can mimic nodular fasciitis, but is relatively slow growing so would be unlikely here.

12) A

Placental alkaline phosphatase is detectable in the blood in **50% of cases of classical seminoma**. Choriocarcinoma is associated with raised hCG, while yolk sac tumour is associated with raised alpha-fetoprotein. Teratocarcinoma refers to a germ cell tumor that is a mixture of teratoma with embryonal carcinoma, or with choriocarcinoma, or with both. Of note is that 10% of seminomas also secrete hCG.

13) E

All of the listed factors except **Down's syndrome** are risk factors for testicular germ cell tumours. Cryptorchidism is probably the most significant. Other more minor risk factors are early onset of male characteristics, the presence of an inguinal hernia, and sedentary lifestyle. The large majority of cases have no identifiable risk factor however, and no definite reason has been identified to explain the doubling of incidence of testicular cancer worldwide since the 1960s.

14) A

Meigs' syndrome is the triad of **ascites, pleural effusion** and benign **ovarian tumour**. The pleural effusion is due to a benign ovarian fibroma, and does not recur if the tumour is removed. (B) is Meige's syndrome which is a type of orofacial dystonia. (C) Refers to Denys-Drash syndrome, a very rare syndrome characterized by the triad of pseudohermaphroditism, mesangial renal sclerosis, and Wilms tumour. (D) is POEMS syndrome, which includes Polyneuropathy, Organomegaly, Endocrinopathy (O)edema, M-protein (an abnormal macroglobulin) and Skin abnormalities (including hyperpigmentation and hypertrichosis). (D) is Gardner syndrome, characterised by multiple colonic polyps together with thyroid carcinomas and epidermal cysts as well as skull osteomas. (E) is Sweet's syndrome or acute febrile neutrophilic dermatosis.

15) E

Alpha-fetoprotein may be raised in all the above tumours except **embryonal carcinoma**, which has no associated tumour marker. Other rare alpha-fetoprotein secreting tumours include the carcinomatous component of a mixed Müllerian tumour, and, on rare occasions, Sertoli-Leydig cell tumour.

16) E

HNPCC, also referred to as Lynch syndrome, is divided into Lynch syndrome I (familial colonic carcinoma) and **Lynch syndrome II**, which also carries an increased risk of other carcinomas, usually **endometrial, ovarian**, urinary tract, small intestinal or **gastric** in origin. Individuals with HNPCC have an approximately 80% lifetime risk of colonic carcinoma, mainly of the right side. These cancers frequently show mucinous differentiation. Inheritance is autosomal dominant. The syndrome is associated with defects in DNA mismatch repair that lead to microsatellite instability, also known as MSI-H, which is a hallmark of the condition.

17) B

Patients with FAP often develop **hepatoblastomas**. Sporadic hepatoblastomas appear to be associated with beta-catenin mutations, which occur in up to 70% of patients.

18) A

Sarcoma botryoides refers to the presentation of **embryonal rhabdomyosarcoma** within the walls of hollow, mucosa lined structures such as the nasopharynx, common bile duct, urinary bladder of infants and young children or the vagina in females. The microscopic appearances suggest "bunches of grapes". Microscopically, cross striations may be visible within cells, confirming them to be rhabdomyoblasts, and a distinct layer of closely packed tumour cells may be visible immediately beneath the mucosal epithelium (a cambium layer). Myo D1 serves as a useful immunohistochemical marker of rhabdomyosarcoma.

19) C

CD117, also known as c-kit, is expressed by 95% of all GISTs. The efficacy of imatinib, a CD117 inhibitor, is determined by the mutation status of CD117. If positive, this now forms an important part of the therapy of GISTs. CD34, although positive in many GISTs, does not influence therapy. CD43 is a T-cell and myeloid marker. Cyclin D1 is positive in mantle cell lymphoma. CD30 is an activation marker which is positive in some lymphomas including anaplastic large cell lymphoma and classical Hodgkin's lymphoma, as well as embryonal carcinoma in some cases.

20) B

Alport syndrome is caused by mutations in COL4A3, COL4A4, and COL4A5; collagen biosynthesis genes. Failure to form normal type IV collagen results in **focal segmental glomerulosclerosis**. Only certain foci of glomeruli within the kidney are affected, and then only a segment of an individual glomerulus. The pathological lesion is sclerosis (fibrosis) within the glomerulus and hyalinisation of the feeding arterioles, but without an increase in the number of cells (hence non-proliferative). The hyaline is an amorphous material, pink, homogeneous, resulting from combination of plasma proteins, increased mesangial matrix and collagen. Staining for antibodies and complement is essentially negative. Steroids are often tried but have not been shown to be effective. 50% of patients with FSGS continue to have progressive deterioration of kidney function, ending in renal failure.

21) E

Membranous glomerulonephritis is usually idiopathic, but may be associated with carcinomas of the lung and bowel, infections including hepatitis and malaria, drugs including penicillamine, and connective tissue diseases such as systemic lupus erythematosus. **Digoxin** is not known to be associated with membranous glomerulonephritis.

22) A

SUDI is defined as cases of deaths in infancy where **no adequate explanation** for death is revealed by a detailed autopsy. Most pathologists use it to mean that whilst no cause has yet been identified and the definition of SIDS cannot yet be met, they have no cause for suspicion and the funeral can therefore proceed, with the death being initially registered as SUDI, with a more precise diagnosis (which may be SIDS) following the full results of investigations.

In a recent study of the causes of SUDI (in 623 cases) the most common diagnosis, when sudden infant death syndrome (SIDS) was excluded, was infection (7.1%), followed by cardiovascular anomaly (2.7%), abuse (2.6%), and metabolic or genetic disorders (2.1%). SIDS is essentially a diagnosis of exclusion therefore. Healthy infants matched for sex, age and season of birth show a distinctly different nasopharyngeal flora to SIDS/SUDI infants, with increased incidence of positive culture for staphylococci, streptococci and gram negative organisms in the latter. Nasal swab is therefore an essential part of investigation. Likewise, culture of a portion of spleen obtained under aseptic conditions is valuable and complements the results of blood culture. Pure growth of a pathogenic bacterium from CSF, associated with an increase in CSF inflammatory cells, is highly likely to be the cause of death. This is as opposed to a situation where mixed organism growth occurs in the absence of increased inflammatory cells, which is likely to be the result of contamination.

23) B
Risk of death from both these syndromes peaks at **2 to 3 months** of age. This correlates with a fall in serum IgG levels and a rise in susceptibility to bacterial infections. Both syndromes are rare after the age of 12 months.

24) A
Group B Streptococcus, via maternal transmission, remains the commonest cause of septicaemia and meningitis in this age group. About half of the cases of group B strep disease among newborns happen in the first week of life, and most of these cases start a few hours after birth. Of the other organisms mentioned, only Neisseria meningitidis and Haemophilus influenzae are significant causes of infant septicaemia and meningitis, and these are considerably less frequent than group B strep in infants under one month.

25) D
Approximately 30 to 40% of cases of **HIV-associated Burkitt's lymphoma** show positivity for EBV. EBV infection is not associated with precursor ALL of either B or T cell types. 90 to 95% of cases of endemic Burkitt's lymphoma are EBV positive. At least 95% of cases of NK/T cell lymphoma of nasal type are positive for EBV.

26) B
Classical Hodgkin's lymphoma accounts for approximately 30 to 40% of all childhood lymphomas, by far the largest category overall. Burkitt's lymphoma, although common in its endemic form in Africa, is relatively rare as a sporadically occurring disease in the West. Nodular lymphocyte predominant Hodgkin's lymphoma is rare in childhood, having a peak incidence in the fourth decade. Lymphoblastic lymphomas are relatively frequent in childhood although less common than Hodgkin's lymphoma. Peripheral T-cell lymphoma is usually a disease of older adults and is exceptionally rare in the paediatric age group.

27) B
This is a rare disease, usually of infants, in which a haemangioma leads to decreased platelet counts (**thrombocytopaenia**).

28) D

A defining feature of anaplastic large cell lymphoma is that it is of T-cell lineage. Another defining feature is positivity for the activation antigen CD30. The latter is not however discriminatory from Hodgkin's lymphoma, of which over 90% of cases are also CD30 positive. There is a high incidence of extranodal presentation in anaplastic large cell lymphoma, with skin, bone marrow and lung frequently involved. However, primary anaplastic large cell lymphoma of the skin must be distinguished from systemic ALCL with secondary cutaneous involvement. Around 75% of cases show the t(2;5) translocation by PCR. Anaplastic large cell lymphoma is **never of B-cell derivation**; some diffuse large B-cell lymphomas do however show an anaplastic morphology that does not have any bearing on prognosis. Additionally, occasional cases of diffuse large B-cell lymphoma may be CD30 positive but this is not of any clinical significance.

29) D

Popcorn, or lymphocytic and histiocytic, cells are characteristic of nodular lymphocyte predominance Hodgkin lymphoma, rather than classical Hodgkin's lymphoma. All the other subtypes are seen in classical Hodgkin's lymphoma. Mummified cells are apoptotic Hodgkin's or Reed-Sternberg cells. Reed-Sternberg cells possess by definition a minimum of two nuclear lobes, usually with prominent eosinophilic nucleoli and amphophilic cytoplasm. Hodgkin cells are the mononuclear variants of Reed-Sternberg cells. Lacunar cells are Hodgkin or Reed- Sternberg cells which appear to be present in a space or lacuna. This is probably an artefact of paraffin processing as it is not seen in frozen sections.

30) D

Metastatic adenocarcinoma of unknown origin is far rarer than previously thought, as improved immunohistochemistry has enabled accurate diagnosis of the primary tumour in the vast majority of cases. It now represents approximately 3 to 4% of all diagnoses of malignancy. Median survival remains poor, with a median survival of around four months. Certain clinical categories have a far better prognosis, however, including metastatic breast carcinoma to axillary lymph nodes and metastatic prostatic carcinoma to bone. Serology for tumour markers forms an important part of investigation. These include CA125 and PSA, **but CA19-9 is insufficiently sensitive and specific to be of value in most cases**. Electron microscopy can be helpful in selected cases in distinguishing carcinoma from sarcoma lymphoma, but cannot reliably differentiate different subtypes of adenocarcinoma.

31) C

Carcinoma of the tongue is probably one of many carcinomas and lymphomas associated with HIV/AIDS. Its increased incidence is believed to be due to a combination of the effects of HPV and EBV on a background of suppression of cell-mediated immunity.

32) E

Basal cell carcinoma (BCC) is the commonest human cancer, with approximately 30% of Caucasian individuals developing a BCC during their lifetimes. Around 80% of cases occur in the head and neck area. Overall, around two thirds of cases are associated with sunlight exposure and damage while approximately one third are not. There is a strong association with basal cell naevus syndrome (Gorlin's syndrome). **Gardner's syndrome is a genetic disorder characterized by the presence of**

multiple polyps in the colon together with tumours outside the colon. The extracolonic tumours may include osteomas of the skull, thyroid cancer, epidermoid cysts, fibromas and sebaceous cysts.

33) C

Gorlin's syndrome is an autosomal dominant condition best known for its association with basal cell carcinoma, but also with, variably, macrocephaly, cleft lip and palate and a variety of skeletal malformations, as well as the presence of odontogenic keratocysts of the jaw. Recent molecular genetic studies have demonstrated Gorlin's syndrome to be caused by mutations in the PTCH (Patched) gene on chromosome arm 9q. **There is no association with colonic carcinoma.**

34) C

Aschoff bodies are seen in the heart in rheumatic fever. They represent areas of interstitial inflammation, showing fibrinoid change, lymphoplasmacytic infiltration, and macrophages (including multinucleated giant cells and so-called caterpillar cells with an abnormal nuclear chromatin pattern) around necrotic centres. Mallory bodies are made up of intermediate filament keratin proteins and are seen in the liver in a variety of conditions of which alcoholic hepatitis is the most important. Howell-Jolly bodies are tiny fragments of residual nuclear material seen in circulating red blood cells. They are not normally present as they are usually removed within the spleen, but are often seen in patients who have had a splenectomy. Psammoma bodies are spherical lamellated particles of calcium associated with papillary thyroid cancer, ovarian papillary serous carcinoma and meningioma. Pick bodies are silver-staining, spherical bodies formed of tau protein in neurons in Pick's disease. Negri bodies are eosinophilic cytoplasmic inclusion bodies found in neurons in rabies, especially in the hippocampus.

35) A

A primary lesion caused by mycobacterium bacilli (tuberculosis) developing in the lung of a previously uninfected individual is called a **Ghon focus**. If it also involves infection of surrounding lymph nodes, it is known as a Ghon complex. A Ghon's focus results from haematogenous seeding at the time of initial infection, is present at the apex of the lung, contains viable organisms and serves as the source of reactivation in approximately 10% of people. The last two terms do not exist.

36) E

All of the above statements are true with the exception that negative **ZN staining cannot exclude Mycobacterium infection**. This is because organisms are frequently present in small numbers in histological and cytological samples. For sputum analysis, fluorescence microscopy is the preferred technique for detecting TB without proceeding to culture. Microbiological culture produces a far higher detection rate than any kind of histochemical staining for mycobacterium bacilli.

37) E

Kikuchi's disease, or histiocytic necrotising lymphadenitis, affects mainly young people with a female preponderance. There may be fever, malaise, headache and lymphadenopathy, with hepatosplenomegaly in a few cases. The disease is generally self-limiting although steroids are sometimes employed in treatment. Patchy areas of necrosis with abundant karyorrhectic debris are characteristic. Aetiology is unknown although viral infection is suspected. **Autoimmune antibodies are not raised in Kikuchi's disease.**

38) **D**

Fas ligand (FasL) and tumour necrosis factor (TNF) are believed to be involved in the initiation of the apoptotic pathway. Proteolytic caspases are part of a group of cysteine proteases, which are involved further downstream in the actual process of enzymatic cellular breakdown. P53 prevents a damaged cell from replicating by stopping the cell cycle at G1 (interphase), to enable the cell to repair itself, however it induces apoptosis if the genetic damage is extensive and repair efforts fail. Disruption to the regulation of the p53 or interferon genes will therefore result in impaired apoptosis and this may result in oncogenesis. **CD103 is an integrin that is a T-cell marker.** It has no function in apoptosis.

39) **C**

The thickness, **rather than the diameter,** of melanomas as expressed by Breslow's thickness, is an important prognostic factor. Other significant prognostic indicators other than those mentioned above include type of melanoma, presence of lymphatic/perineural invasion, depth relative to skin structures (Clark's level), presence of satellite lesions, and, of course, the presence of distant metastases.

40) **E**

Although frozen sections may be ready in as little as 10 minutes, **technical quality of the sections always remains inferior to paraffin processed ones.** For all practical purposes therefore, only two main questions may be answered: is the tissue "benign" or "malignant", or is a surgical margin or node clear of cancer. Although the use of frozen sections in diagnostic pathology has reduced markedly over the past 40 years, they remain of particular value in Moh's micrographic surgery as well as in demonstrating or excluding metastases in sentinel lymph nodes. Oil red O stain can show fats on frozen but not paraffin processed sections.

41) **C**

Oil Red O is a diazo histochemical stain used for staining lipids. It needs to be performed on fresh or frozen tissue, as paraffin processing removes the lipids from specimens. Congo and Sirius Reds are stains for amyloid. Safranin is principally a counterstain, classically used in Gram staining. Methyl green pyronin was formerly used for demonstrating plasma cells, but has largely been supplanted by immunohistochemistry for CD138.

42) **C**

Silver staining is the classic technique for demonstrating fungal organisms in histological sections, and is usually considered superior to PAS staining for this purpose. Bismarck brown staining is for the detection of acid mucins. Crystal violet is used as part of Gram staining. Sudan black is a stain for lipids. Ziehl-Neelsen stain is used to detect acid-fast organisms.

43) **B**

Karyorrhexis is nuclear fragmentation as part of either apoptosis or necrosis. It is usually preceded by pyknosis, which is nuclear condensation, and followed by karyolysis, which is complete dissolution of chromatin. Apoptosis is the process of programmed cell death whereas necrosis is premature cellular death.

44) B

The benign adaptive change that occurs in epithelia in response to chronic irritation or other stresses is called **metaplasia**; in principle it is reversible. It is exemplified by the transformation of respiratory type to squamous epithelium in the upper respiratory tract caused by smoking. Dysplasia is a precancerous condition characterised by maturation and differentiation abnormalities. Hypertrophy is an adaptive process involving increase in cell size (e.g. cardiac hypertrophy in athletes), while hyperplasia is an often physiological, non-neoplastic, proliferation of cells in response to a stimulus (e.g. endometrial hyperplasia). Anaplasia is a complete loss of differentiation in malignant tumours.

45) B

Glutaraldehyde is the classic fixative for electronmicroscopy. Very small/thin portions of tissue must be used as its penetrative abilities are poor. B5 and Zenker's fluids are mercury-based fixatives that are sometimes used for bone marrow and other lymphoreticular specimens as they produce excellent nuclear detail. Bouin's fluid has traditionally been used for testicular biopsies and small lymph node biopsies, as it also gives better nuclear detail than neutral buffered formalin. It has the additional advantage that it colours tiny biopsies yellow and makes them easier to see and more difficult to lose. Neutral buffered formalin is of course the standard fixative used in most histopathological processing.

46) B

Giemsa stain, as well as being used for peripheral blood smears and bone marrow aspirates, is useful for highlighting microorganisms including histoplasma and malaria. Warthin-Starry stain is used for the detection of spirochetes such as Treponema and can also detect Helicobacter. Silver stain can detect fungal organisms such as pneumocystis. Van Gieson's Stain is used for staining collagen but is not helpful in detecting microorganisms. The main appearance of methylene blue in the histopathology lab is its use as a stain by endoscopists to identify dysplastic mucosa or to identify sentinel lymph nodes. Ironically it is also a cheap and fairly effective treatment for malaria, although best used in combination with other drugs.

47) A

Clonorchis sinensis is a liver fluke that inhabits the biliary tree (where it feeds on bile) and infects up to 30,000,000 people in Southeast Asia and Japan. It is spread by the ingestion of infected fish. It engenders chronic inflammation of the biliary tree that predisposes to the development of cholangiocarcinoma. Paragonimus is a lung fluke that is also widespread in Southeast Asia and believed to be the commonest cause of haemoptysis worldwide. Ascaris may enter the biliary tree but is more classically associated with the gastrointestinal and respiratory tracts. Dracunculus, or guinea worm, usually remains confined to the subcutaneous tissues. Taenia saginata is the beef tapeworm and normally dwells in the lumen of the gastrointestinal tract.

48) C

Löffler's syndrome is classically associated with infection by **Ascaris lumbricoides**. It is an eosinophilic pneumonia that manifests with cough, breathlessness and fever.

Echinococcus granulosus is the cause of hydatid disease, with formation of cysts in multiple organs, especially the liver. Onchocerca is a nematode that is the cause of river blindness in Africa. Loa loa is another nematode that causes lymphatic filariasis and conjunctival inflammation. Toxoplasma is a protozoan that causes a usually self-limiting, sometimes flu-like, illness in healthy people but may be fatal in the immunosuppressed or in fetuses.

49) **B**
Histologically, toxoplasmosis is characterised by the triad of follicular hyperplasia, small, poorly formed epithelioid granulomas, and monocytoid B-cell hyperplasia. **Necrotising granulomatous inflammation is not a feature.** The disease is geographically widespread, with approximately one third of the world's population having been infected and carrying the disease in the form of latent toxoplasmosis, where the disease (bradyzoites) remains present as tiny cysts in muscle and nerves. A number of recent studies have suggested that toxoplasmosis may underlie a variety of neurological problems ranging from slow reaction time to schizophrenia. The disease is usually clinically asymptomatic and apart from a flulike illness in some people is likely to pass unrecognised. Most fatalities occur in the immunocompromised and in fetuses as a result of fetal-maternal transmission.

50) **B**
50 to 60% of all cases of hypertensive brain haemorrhage affect the **putamen**. Less common sites include the other deep white matter structures and the white matter of the cerebrum. The cerebellum is a rare site for hypertensive haemorrhages.

51) **E**
Malignant **rhabdoid tumour** (MRT) is a very aggressive form of tumour originally described as a variant of Wilms tumour, which is a kidney tumour that occurs mainly in children. Histologic diagnosis depends on identification of characteristic rhabdoid cells—large cells with eccentrically located nuclei and abundant, eosinophilic cytoplasm.

52) **A**
Granulocytic (or myeloid) sarcoma refers to a tissue deposit of acute myeloid leukaemia; it is not a sarcoma. In most cases circulating myeloid blasts are present in peripheral blood.

53) **B**
Alveolar rhabdomyosarcoma, which resembles pulmonary alveoli, is associated with a fusion protein between PAX3 and FKHR (now known as FOXO1). It commonly occurs in older children and teenagers.

54) **D**
Ewing sarcoma is caused by a translocation between chromosomes 11 and 22, which fuses the EWS gene on chromosome 22 to the FLI1 gene on chromosome 11. It is found preferentially in bony sites such as the pelvis, femur, humerus and ribs.

55) **C**

Embryonal rhabdomyosarcoma is composed of small, round tumour cells and some large, polygonal cells with eosinophilic cytoplasm, which may contain diagnostic cross striations. The cells have a similar appearance to embryo cells aged 6 – 8 weeks.

56) B

Synovial sarcoma may be monophasic, purely composed of spindle cells, or can be biphasic, with epithelioid areas also. Generous sampling is important to identify the latter. Synovial sarcoma is usually positive to immunohistochemical challenge for **broad spectrum cytokeratins**.

57) A

Malignant peripheral nerve sheath tumour, MPNST, may be sporadic or may occur secondary to neurofibromatosis. Immunopositivity for neural markers (**S100**, NF, **NSE**, myelin basic protein) is usually present.

58) E

Malignant fibrous histiocytoma, MFH, should only be used as a categorisation of last resort, if it is not possible to assign a sarcoma to a specific lineage by detailed immunohistochemical study. This is sometimes referred to as a "**dustbin diagnosis**."

59) C

This rare sarcoma was previously known as **inflammatory pseudotumour**. It is commoner in women, who often present with pyrexia of unknown origin, PUO, or other nonspecific findings.

60) D

Liposarcoma is characterised by the presence of lipoblasts but these are not in themselves diagnostic; close clinical and radiological correlation is needed as deeper seated lesions are much more likely to be malignant. The lipoblasts are often S100 positive.

61) C

Acinic cell carcinoma is a rare tumour mainly occurring in the parotid gland but which may also occur in the breast. This is an adenocarcinoma with demonstrably vacuolated cells showing acinar differentiation and DPAS positive granules. Unlike normal serous tissue there are no ducts or lobular organization.

62) E

Adenoid cystic carcinoma may also arise in breast or skin. It is fairly indolent, with disease specific survival at 5 yrs of 89%. It is notorious for its perineural spread. The tumour is composed of small basaloid myoepithelial cells that are usually SMA positive.

63) A

Polymorphous low grade adenocarcinoma demonstrates bland monomorphic cells and usually arises from minor salivary glands. As the name suggests, *polymorphous* low-grade adenocarcinoma may show many patterns of differentiation. It usually arises in the palate.

64) D

The question stem describes an **epithelial-myoepithelial carcinoma**. This rare tumour makes up only 0.2% of salivary gland carcinomas. It is low grade and usually occurs in the parotid gland. It tends to occur in older women.

65) **B**

Lymphoepithelial carcinoma is actually a nasopharyngeal carcinoma; essentially a very poorly differentiated nonkeratinising squamous cell carcinoma. It is common in China and is associated with EBV infection.

66) **A**

Granulosa cell tumours usually occur in postmenopausal women. The presentation with postmenopausal bleeding occurs because of secretion of oestrogen by the tumour.

67) **D**

Sertoli cell tumours of the ovary are rare. They show a tubular pattern and typically present in young females, who may undergo virilization.

68) **C**

Fibromas are firm pale tumours. A variant with ascites and pleural effusion may occur Meig's syndrome. On microscopic examination, there are intersecting bundles of spindle cells producing collagen. Mitoses are absent or rare.

69) **I**

The question stem describes the typical features of a **fibrosarcoma**.

70) **E**

The question stem describes a **Leydig cell tumour**. Reinke's crystals may be present. The prognosis is generally good, as the tumours tend to be benign. Around 10% are malignant.

71) **C**

In **Lewy body** dementia there is a loss of dopamine-producing neurons (in the substantia nigra) similar to that seen in Parkinson's disease.

72) **D**

The involuntary movements described in the question stem are termed dance-like or "chorea". These movements are part of **Huntington's chorea** that is inherited in an autosomal dominant fashion.

73) **E**

Primary CNS lymphoma is recognized as a complication in individuals infected with HIV. It is usually a diffuse large B cell lymphoma and carries a poor prognosis. Additionally this lymphoma is increasing in incidence in the immunocompetent individuals of the general population.

74) **A**

Ataxia, ophthalmoplegia, confusion and impairment of short-term memory are all features of **Wernicke's encephalopathy**.

75) **G**

The distinct spherical nodules that can be discerned macroscopically and are causing oedema are likely to be metastases. The Masson Fontana staining is consistent with the presence of melanin. If S100, MelanA, HMB45 or NK1C3 immunohistochemical challenge on the lesion had been carried out, they would have confirmed the presence of melanocytes in the nodules. The deceased patient was likely to have suffered from **metastatic malignant melanoma**.

76) **A**

AE1/3 is a broad-spectrum cytokeratin that stains most **carcinomas** but is not diagnostically specific.

77) **B**

CAM 5.2 is a low-molecular weight cytokeratin that stains most **adenocarcinomas**, but not squamous carcinomas.

78) **E**

A CK 5/6 cocktail is normally used for **squamous cell carcinomas**.

79) **C**

The majority of cases of **TCC** are CK20 positive, CK7 negative, which may help in differentiating them from some adenocarcinomas if poorly differentiated.

80) **D**

TTF-1 is positive in lung and thyroid adenocarcinomas and is used for monitoring for metastasis and recurrence.

81) **A**

GCDFP-15 stains **breast carcinoma**, salivary duct carcinoma and apocrine epithelium.

82) **C**

Mesothelin is overexpressed in several human tumours, including **mesothelioma**, ovarian adenocarcinoma and pancreatic adenocarcinoma.

83) **B**

CDX2 is used in diagnostic surgical pathology as a marker for gastrointestinal differentiation, especially **colorectal**.

84) **E**

Clear cell renal cell carcinoma is CK20 negative, CK7 negative in the vast majority of cases.

85) **F**

Most **pancreatic adenocarcinomas** are CK20 and CK7 positive.

86) **B**

The question stem describes the immunohistochemical profile of **chronic lymphocytic leukaemia/small lymphocytic lymphoma**. Histologically, the presence of proliferation centres can also be useful diagnostically.

87) **A**

The question stem describes the immunohistochemical profile of **marginal zone lymphoma**. Histologically, follicular colonisation and the presence of monocytoid and/or plasmacytoid cells are also helpful diagnostically.

88) **C**

The question stem describes the immunohistochemical profile of **mantle cell lymphoma**. A high degree of monomorphism and the presence of eosinophilic histiocytes are also characteristic of this lymphoma.

89) **D**

The question stem describes the immunohistochemical profile of **follicular lymphoma**. Only 60-70% of follicular lymphomas are CD10 positive but the follicular architecture is seen at least focally in nearly all cases.

90) **E**

The question stem describes the immunohistochemical profile of **acute lymphoblastic leukaemia, ALL**. Tdt is positive in almost all cases of ALL. CD10 and CD34 are also helpful markers.

91) **D**

The question stem describes **Tangier disease**. Tangier disease is a rare disorder with approximately 50 cases reported worldwide. The decreased serum levels of HDL and tissue accumulation of cholesterol predispose to early atherosclerosis, hepatomegaly and splenomegaly.

92) **F**

The question stem describes **Rotor syndrome** – a cause of conjugated hyperbilirubinaemia due to an intrahepatic metabolic defect. Rotor type hyperbilirubinaemia is a rare, relatively benign autosomal recessive bilirubin disorder of unknown origin.

93) **E**

The question stem describes **X-linked ichthyosis**. XLI, also known as X-linked recessive ichthyosis, manifests with dry, scaly skin and is due to deletions or mutations in the STS (steroid sulfatase) gene.

94) **G**

The question stem describes **Wolman disease**. This rare autosomal recessive disorder is a lysosomal storage disease. Mutations in the *LIPA* gene cause Wolman disease. The *LIPA* gene provides instructions for producing an enzyme called lysosomal acid lipase. This enzyme is found in the lysosomes (compartments that digest and recycle materials in the cell), where it processes lipids such as cholesteryl esters and triglycerides so they can be used by the body. Almost all sufferers of Wolman disease die before the age of one.

95) **B**

The question stem describes **Menkes disease**. Menkes disease, also known as kinky hair disease, is an X-linked neurodegenerative disease of impaired copper transport.

Menkes disease typically manifests during infancy with hypotonia, seizures, mental retardation, and developmental delay.

96) E

Option (E) is a description of **Parkinson's disease**. The implied observable change in the substantia nigra is depigmentation. The loss of dopaminergic neurons is the underlying primary pathology in Parkinson's disease.

97) A

Alzheimer's disease is the most common form of dementia, characterised by loss of neurons and synapses in the cerebral cortex and certain subcortical regions. Amyloid plaques and neurofibrillary tangles occur together in Alzheimer's disease. Alzheimer's disease is the commonest cause of dementia in the world.

98) D

The pathognomonic body is Pick's body that occurs in **Pick's disease**. 3R tau protein is the dominant tau protein present in these bodies. Pick's disease causes progressive neuronal destruction and causes the tau proteins in neurons to aggregate into the silver staining, spherical Pick bodies.

99) B

Huntington's disease is a hereditary disease that is one of a group of trinucleotide repeat disorders. The disease can show anticipation with each generation and usually presents with a loss of coordination and unsteady gait. The HTT gene is on the short arm of chromosome 4.

100) C

Motor neurone disease, MND, is a progressive neurodenegenerative disease that in its most severe cases can progress to paralysis, loss of speech and difficulty swallowing. Cases of Motor Neurone Disease, MND, show degeneration of the ventral horns of the spinal cord, as well as atrophy of the ventral roots. In the brain, atrophy may also be present in the frontal and temporal lobes.

(F) is a description of Wilson's disease.
(G) is a description of Neurosyphilis.
(H) is a description of Herpes simplex encephalitis.

101) E

Melanosis coli causes the deposition of **pseudomelanin** in colorectal mucosa. Pseudomelanin is a PAS positive pigment. Melanosis coli is a benign condition related to longterm laxative use, with little pathological significance.

102) A

Xanthelasma represent macroscopically observable subcutaneous deposits of fat, usually cholesterol, around the eyelids. Many cases of xanthelasma occur in healthy individuals, but there is an increased incidence of hyperlipidaemia or hypercholesterolaemia in people who have them.

103) J

Pompe's disease is a lysosomal **glycogen** storage disease that is due to a hereditary deficiency of the enzyme α-1,4-glucosidase.

104) C
Niemann-Pick's disease is one of the inherited lysosomal storage diseases; it causes **sphingomyelin** accumulation in multiple organ systems. Type A tends to cause manifestations in the nervous system whereas type B has no central nervous system involvement, instead often causing splenic enlargement.

105) D
Mallory's hyaline is a recognized feature of alcoholic liver disease, Wilson's disease, primary biliary cirrhosis, other non-alcoholic cirrhosis, hepatocellular carcinoma and severe obesity. Mallory's hyaline is composed of intermediate keratin filaments.

106) A
Option (A) describes **Lambert-Eaton myasthenic syndrome**. This is a disease of the neuromuscular junction, which is thought to be caused by an autoimmune attack on the voltage gated calcium channels on the presynaptic cell membrane. In contrast to myasthenia gravis, which it may resemble, symptoms of Lambert-Eaton myasthenic syndrome, LEMS, tend to be worse in the morning and improve with exercise and nerve stimulation.

107) C
Option (C) describes key features of **myasthenia gravis**. Myasthenia gravis is a disease of the neuromuscular junction caused by an autoimmune attack on the acetylcholine receptor. Initial symptoms may include generalized muscle weakness that fluctuates and responds well to anticholinesterases. Management of the disease can involve the use of immunosuppressants or thymectomy.

108) E
Option (E) describes **Guillain-Barré syndrome** ("acute inflammatory demyelinating polyradiculoneuropathy"). Guillain-Barré syndrome is associated with a range of viral or bacterial infections and is believed to have an underlying autoimmune origin. This disease causes a peripheral neuropathy that has a glove and stocking distribution initially, which progresses as an ascending paralysis that can be life-threatening. Treatment for Guillan-Barre syndrome involves plasmapheresis or intravenous immunoglobulins and supportive care, the majority of patients making a full recovery.

109) D
Option (D) describes **Fragile X syndrome**. The underlying pathology is a repeating trinucleotide sequence. Fragile X syndrome results in a failure to express the protein coded by the FMR1 gene, necessary for normal neural development.

110) F
Option (F) describes **multiple sclerosis**. This is the commonest demyelinating disease and is believed to have an autoimmune origin. It is characterized by episodes of neurological deficit separated in time and space. Optic neuritis is a common presentation of multiple sclerosis.

> **(B)** This is a description of Lyme disease.
> **(G)** This is a description of diabetes mellitus.
> **(H)** This is a description of Duchenne muscular dystrophy.

111) B

Option (B) describes **bronchioloalveolar carcinoma** (BAC). This is essentially a very well differentiated adenocarcinoma that exists in mucinous or non-mucinous subtypes, arising in the distal bronchioles or alveoli. It grows along the pre-existing lung structure without destroying alveolar architecture. Histologically there is usually no evidence of stromal, vascular or pleural invasion. Macroscopically and radiologically BAC was often confused with pneumonic consolidation.

112) D

Option (D) describes a **carcinoid tumour**. Carcinoids are neuroendocrine tumours. Carcinoids of the lung are regarded as low-grade malignant neoplasms because they are locally invasive, may undergo local recurrence, and occasionally metastasise.

113) E

Option (E) describes **small cell carcinoma**. Both small cell carcinomas and carcinoids are neuroendocrine tumours. Small cell carcinoma is a highly malignant tumour thought to derive from neuroendocrine cells (APUD cells) in the bronchus.

114) C

Option (C) is a classical description of a primary **adenocarcinoma** of the lung. In contrast option (G) describes a specific subset of adenocarcinoma, the metastatic renal cell adenocarcinoma.

115) A

Option (A) describes **large cell carcinoma** of the lung. This tumour comprises 5%-10% of all lung cancers. Large cell carcinoma is a diagnosis of exclusion, having ruled out better-differentiated cancers (squamous cell carcinoma, adenocarcinoma and small cell carcinoma). Option (F) describes squamous cell carcinoma.

116) C

Ferruginous bodies refer to asbestos bodies coated in particles of iron, giving a beaded or "dumbbell" appearance.

117) B

Homogentisic acid accumulates in ochronosis. The dark colour accounts for its alternative name of "melanic acid".

118) A

Keloids are formed of **type III collagen** deposited in excessive amounts.

119) D

Mott cells are plasma cells containing multiple **immunoglobulin** inclusions.

120) E

Fibrin is the major constituent of the hyaline membranes in both infant and adult respiratory distress syndrome.

121) **J**

The question stem description is consistent with a **pulmonary hamartoma**. This is a relatively common benign neoplasm composed of cartilage, connective tissue, muscle, fat, and bone.

122) **B**

The question stem describes a **Krukenberg tumour**. Comparison of the morphology of the primary and Krukenberg tumour, confirms the origin of the tumour.

123) **E**

Masson's tumour is synonymous with intravascular papillary endothelial hyperplasia. Its commonest sites of occurrence are in the head and neck region.

124) **F**

The question stem describes an **eccrine poroma**. The tumour can sometimes be invasive – then an invasive edge, pleomorphism and mitoses are prominent.

125) **H**

The question stem describes **linitis plastica**; the desmoplastic response to the diffuse infiltration of a poorly differentiated adenocarcinoma causes hardening of the gastric wall.

PAPER 4

1) Which of the following tumours of the respiratory tract is defined by CD34 positivity?

(A) Mesothelioma
✓(B) Solitary fibrous tumour of the pleura
(C) Small cell carcinoma
(D) Squamous cell carcinoma
(E) Carcinoid tumour

2) The following infections may occur at any time during gestation or occasionally at the time of delivery, via maternal-to-fetal transfusion, except:

(A) Toxoplasmosis
✓(B) Infectious mononucleosis
(C) Rubella
(D) Cytomegalovirus
(E) Herpes simplex virus

3) Which of the following conditions is not a vasculopathy?

(A) Atherosclerosis
(B) Monckeberg's medial calcific sclerosis
(C) Leukocytoclastic vasculitis
(D) Buerger's disease
✓(E) Farber's disease

4) Identify the salivary gland tumour that is a papillary adenoma composed of eosinophilic, mitochondrion-packed oncocytes ("Hürthle cells") with a dense stromal lymphocytic infiltrate.

✓(A) Warthin's tumour
(B) Pleomorphic adenoma
(C) Adenoid cystic carcinoma
(D) Clear cell carcinoma
(E) Mucoepidermoid carcinoma

5) The following statements are true of metaplasia, except:

(A) The most common form of metaplasia is from columnar to squamous epithelium.
✓(B) It does not occur in connective tissue.
(C) It can predispose to malignant transformation.
(D) Barrett's or columnar lined oesophagus is a form of metaplasia.
(E) It is mediated by cytokines.

6) The commonest causes of hypercalcaemia include all the following, except

(A) Hyperparathyroidism
(B) Renal failure

(C) Malignant disorders affecting the skeletal system
(D) Vitamin D hypervitaminosis
√(E) Acute pancreatitis

7) Which of the following tumours is a low-grade malignancy that most commonly occurs in patients under 40 and represents less than 5% of all lung tumours?

√(A) Inflammatory myofibroblastic tumour
(B) Hamartoma
(C) Small cell carcinoma
— (D) Carcinoid tumour
(E) Bronchioloalveolar carcinoma

8) The following subtype of lung carcinoma is only very rarely associated with paraneoplastic syndromes:

(A) Carcinoid tumour
√(B) Small cell carcinoma
(C) Squamous cell carcinoma
— (D) Adenocarcinoma
(E) Atypical carcinoid tumour

9) Lambert-Eaton myasthenic syndrome is associated with which lung tumour:

(A) Squamous cell carcinoma
(B) Large cell carcinoma
(C) Small cell carcinoma
√ (D) Carcinoid tumour
(E) Pancoast tumour

10) Horner's syndrome is associated with which lung tumour:

(A) Squamous cell carcinoma
(B) Large cell carcinoma
(C) Small cell carcinoma
(D) Carcinoid tumour
√ (E) Pancoast tumour

11) Which of the following disorders is most likely to be associated with an exudate?

(A) Congestive cardiac failure
(B) Cirrhosis
(C) Nephrotic syndrome
√(D) Rheumatoid disease
(E) Chronic renal failure

12) Identify the TNM stage that describes a lung carcinoma involving the chest wall, with metastases to ipsilateral mediastinal lymph nodes but without distant metastases:

(A) T1,N1,M0

(B) T3,N1,M0
(C) T4,N1,M1
(D) T2,N2,M0
✓(E) T3,N2,M0

13) Which of the following types of primary lung tumour has the weakest association with tobacco use?

(A) Adenocarcinoma
(B) Squamous cell carcinoma
(C) Malignant mesothelioma
(D) Large cell carcinoma
✓(E) Bronchioloalveolar carcinoma

14) Which of the following is not an important cause of pneumonia in immunocompromised patients?

(A) Cytomegalovirus
(B) Pneumocystis carinii
(C) Aspergillus
(D) Cryptococcus
✓(E) Borrelia burgdorferi

15) Identify the organism that is an accepted cause of atypical pneumonia:

(A) Staphylococcus aureus
(B) Klebsiella pneumoniae
(C) Pseudomonas aeruginosa
✓(D) Mycoplasma pneumoniae
(E) Streptococcus pneumoniae

16) The following factors increase the risk of reflux oesophagitis, except:

(A) A sliding hiatus hernia
(B) Helicobacter pylori infection
(C) Smoking
(D) Excess alcohol consumption
✓(E) Foods rich in tomatoes

17) Which of the following increases the risk of oesophageal adenocarcinoma?

✓(A) Long-standing reflux oesophagitis
(B) Coeliac disease
(C) Plummer-Vinson syndrome
(D) Ectodermal dysplasia
(E) Epidermolysis bullosa

18) Identify the genetic condition that is associated with an increased incidence of congenital hypertrophic pyloric stenosis:
(A) Down's syndrome

(B) Hunter's syndrome
(C) Hurler's syndrome
(D) Turner's syndrome
(E) Di George syndrome

19) Which of the following conditions is not associated with Helicobacter pylori infection?

(A) Marginal zone lymphoma
(B) Follicular gastritis
(C) Gastric carcinoma
(D) Hypertrophic gastropathy
(E) Peptic ulceration of the duodenum

20) Which of the following gastric tumours is associated with trisomy 3 and t(11;18)?

(A) Marginal zone lymphoma
(B) Gastrointestinal stromal tumour
(C) Gastric adenocarcinoma
(D) Linitis plastica
(E) Gastric carcinoid tumour

21) All of the following are important causes of bacterial enteritis, except one. Identify the exception:

(A) Vibrio cholera
(B) Campylobacter species
(C) Bifidobacterium species
(D) Mycobacterium tuberculosis
(E) Yersinia enterocolitica

22) Select the following pathological feature characteristically seen in ulcerative colitis:

(A) Skip lesions
(B) Transmural inflammation
(C) Backwash ileitis
(D) Pseudopolyp formation
(E) Fissuring ulceration

23) Squamous cell carcinoma of the cervix involving the pelvic wall and lower third of the vagina implies which FIGO stage?

(A) 1a2
(B) 1b
(C) II
(D) III
(E) IV

24) Which of the following human papillomavirus types has a high risk of generating squamous cell carcinoma of the uterine cervix?

(A) 6
(B) 11
(C) 16
(D) 62
(E) 44

25) Identify the sex cord-stromal tumour of the ovary:

(A) Brenner tumour
(B) Dysgerminoma
(C) Gynandroblastoma
(D) Yolk sac tumour
(E) Struma ovarii

26) Ectopic pregnancies occur at a rate of approximately 1 in how many pregnancies?

(A) 50
(B) 250
(C) 500
(D) 1000
(E) 1500

27) Which of the following conditions refers to the adherence of the placenta directly to the myometrium?

(A) Placenta praevia
(B) Placenta accreta
(C) Placental abruption
(D) Placenta profundus
(E) Placenta membranacea

28) Identify the thyroid malignancy that has an excellent prognosis even if metastasis to local lymph nodes has occurred:

(A) Papillary carcinoma
(B) Follicular carcinoma
(C) Anaplastic carcinoma
(D) Medullary carcinoma
(E) Thyroid sarcoma

29) Which of the following thyroid malignancies is associated with MEN 2A or 2B syndrome in 10 to 20% of cases?

(A) Papillary carcinoma
(B) Follicular carcinoma
(C) Anaplastic carcinoma
(D) Medullary carcinoma

(E) Diffuse large B cell lymphoma

30) Which of the following thyroid malignancies is associated with mutations of the RET proto-oncogene?

(A) Papillary carcinoma
(B) Follicular carcinoma
(C) Anaplastic carcinoma
(D) Medullary carcinoma
(E) Marginal zone lymphoma

31) Primary parathyroid gland hyperplasia accounts for approximately what percentage of cases of primary hyperparathyroidism?

(A) 75-80%
(B) 10-15%
(C) <5%
(D) 50%
(E) >90%

32) Select the disorder that is not autosomal recessive:

(A) Cystic fibrosis
(B) Polycystic kidney disease
(C) Sickle cell anaemia
(D) Haemochromatosis
(E) Wilson's disease

33) Which of the following inborn errors of metabolism results in the accumulation of homogentisic acid, imparting a black colour to urine?

(A) Niemann-Pick disease
(B) Gaucher disease
(C) Galactosaemia
(D) Alkaptonuria
(E) Phenylketonuria

34) Which of the following chromosomal disorders is associated with a high incidence of extragonadal germ cell tumours as well as autoimmune disorders?

(A) Trisomy 21
(B) Trisomy 18
(C) Trisomy 13
(D) Klinefelter's syndrome
(E) Turner's syndrome

35) Which of the following skin conditions is caused by a spirochaete?

(A) Verruca vulgaris
(B) Condyloma acuminatum

(C) Condyloma latum
(D) Molluscum contagiosum
(E) Verruca plana

36) Identify the bone lesion, which characteristically produces severe nocturnal pain that is greatly relieved by aspirin:

(A) Osteosarcoma
(B) Osteoid osteoma
(C) Osteoblastoma
(D) Chondroblastoma
(E) Giant cell tumour

37) Select which of the following bone tumours characteristically arises around the knee joint and gives rise to a soap bubble appearance on plain film x-ray. The vast majority of such lesions do not recur but some may be locally aggressive and a small percentage (under 5%) may give rise to metastases.

(A) Osteoblastoma
(B) Chondroblastoma
(C) Giant cell tumour
(D) Ewing's sarcoma
(E) Non-ossifying fibroma

38) Which of the following is a benign bone tumour, which includes irregular curvilinear bone trabecula set in a fibrous stroma? It occurs in monostotic and polyostotic forms.

(A) Fibrous dysplasia
(B) Nonossifying fibroma
(C) Chondroblastoma
(D) Osteochondroma
(E) Osteoblastoma

39) Identify and select the hepatic lesion that is a well circumscribed nodule with a central stellate scar, most often occurring in young female adults:

(A) Hepatic adenoma
(B) Nodular regenerative hyperplasia
(C) Focal nodular hyperplasia
(D) Haemangioma
(E) Fibrolamellar carcinoma

40) Which of the following is a toxic agent, previously used as intravenous contrast in radiology, is associated with the development of angiosarcoma of the liver?

(A) Thorotrast
(B) Diatrizoate
(C) Iothalamate
(D) Metrizoate

(E) Biligrafin

41) Identify the fungus responsible for producing dietary aflatoxins, which are responsible for a high percentage of the world's cases of hepatocellular carcinoma:

(A) Cyathus stercoreus
(B) Boletus edulis
(C) Amanita muscaria
(D) Aspergillus niger
(E) Aspergillus flavus

42) Which of the following is not a significant aetiological or risk factor for hepatocellular carcinoma?

(A) Haemochromatosis
(B) Hepatic venoocclusive disease
(C) Alcoholism
(D) Cigarette smoking
(E) Hepatitis C infection

43) Causes of post-hepatic jaundice include all of the following except:

(A) Pancreatic cancer
(B) Gallstones in the common bile duct
(C) Carcinoma of the common bile duct
(D) Liver fluke infestation
(E) Gilbert syndrome

44) Which of the following descriptions of abnormalities applies most accurately to post-hepatic jaundice?

(A) Normal urine colour, pale stools
(B) Dark urine colour, pale stools
(C) Absence of conjugated bilirubin in urine
(D) Increased unconjugated bilirubin
(E) Normal alkaline phosphatase, increased aspartate transferase

45) Identify the commonest cause of chronic liver failure:

(A) Alcoholic cirrhosis
(B) Chronic viral hepatitis
(C) Steatohepatitis
(D) Wilson's disease
(E) Haemochromatosis

46) Each of the following is an important cause of acute liver failure (liver failure developing in under 28 days), except:

(A) Paracetamol overdose
(B) Alcoholic hepatitis

148

(C) Hepatitis A
(D) Hepatitis B
(E) Hepatitis C

47) Reye's syndrome is a potentially fatal disease resulting in acute fatty liver and encephalopathy. It is associated with a reaction to which of the following drugs in children?

(A) Paracetamol
(B) Ibuprofen
(C) Diclofenac
(D) Aspirin
(E) Tetracyclines

48) Each of the following signs is characteristic of hepatic encephalopathy on physical examination, except:

(A) Asterixis
(B) Peliosis hepatis
(C) Exaggerated tendon reflexes
(D) Abnormal plantar reflex
(E) Foetor hepaticus

49) What is the commonest congenital abnormality of the pancreas?

(A) Agenesis
(B) Ectopia of pancreatic tissue within the stomach
(C) Pancreas divisum
(D) Annular pancreas
(E) Ectopia of pancreatic tissue within Meckel's diverticulum

50) What is the commonest congenital abnormality of the kidneys?

(A) Unilateral agenesis
(B) Bilateral agenesis
(C) Hypoplasia
(D) Horseshoe kidney
(E) Ectopic kidney tissue within the abdomen

Match each of the following descriptions with the genetic disorder that best fits it:

A) Zellweger's syndrome
B) Phenylketonuria
C) Krabbe disease
D) Hunter syndrome
E) McArdle disease

F) Hermaphroditism
G) Sandhof disease
H) α_1 antitrypsin deficiency
I) Hurler syndrome
J) Cystic fibrosis

51) This is an X-linked recessive disease that causes mucopolysaccharide deposition. There is deficient L-iduronosulphate sulphatase function. Early symptoms include abdominal hernias, ear infections and repeated colds.

52) Also known as cerebrohepatorenal syndrome, results from the inability to oxidize long chain fatty acids in the peroxisomes, causing a spectrum of neurological and ocular problems.

53) This syndrome has a major effect on the metabolism of skeletal muscle. The deficiency is in muscle phosphorylase.

54) This disease inhibits the release of pancreatic secretions and predisposes the sufferer to chronic lung infections.

55) The classic form of this disease is caused by the absence or deficiency of phenylalanine hydroxylase.

Match the following immunohistochemical phenotypic patterns to the most appropriate small to medium cell non-Hodgkin's lymphoma:

A) Burkitt's lymphoma
B) Lymphoplasmacytic lymphoma
C) T cell prolymphocytic leukaemia
D) NK/T cell lymphoma of nasal type
E) Precursor T cell acute lymphoblastic leukaemia/lymphoma

F) Mantle zone lymphoma
G) Follicular lymphoma
H) Marginal zone lymphoma
I) MALT lymphoma

56) CD20 positive, CD5 positive, CD3 positive, CD2 positive, CD4 positive, CD8 negative.

57) CD20 negative, CD3 negative, CD56 positive, CD8 positive, CD4 negative, EBV positive.

58) CD20 negative, CD3 positive, CD34 positive, TdT positive, cyclin D1 negative, CD10 positive.

59) CD20 positive, CD5 negative, CD43 negative, CD23 negative, cyclin D1 negative, CD10 negative, IgM paraprotein detected.

60) CD20 positive, CD5 negative, CD43 negative, bcl-2 negative, cyclin D1 negative, CD10 positive, proliferative index ~100%.

Match the central nervous system pathology below with the most appropriate description:

A) Hepatic encephalopathy
B) Central pontine myelinolysis
C) Vascular dementia
D) Carcinoma metastatic to the CNS
E) Herpes encephalitis
F) Coup injury

G) Contrecoup injury
H) Extradural/epidural haemorrhage
I) Subdural haemorrhage
J) Subarachnoid haemorrhage
K) Bacterial meningitis
L) Intracerebral haemorrhage

61) The strongest risk factor for this type of haemorrhage is systemic hypertension. It is associated with Charcot-Bouchard microaneurysms.

62) Rapid correction of hyponatraemia is the classic cause of this disorder. Even with treatment, this acute condition is fatal in around 20% of patients, with a high incidence of serious long-term neurological sequelae.

63) A 45 year old male adult, who was previously well, has new onset seizures, headaches and changes in behaviour. He works outside on a building site and keeps reaching to scratch something on his back.

64) This haemorrhage is most frequently caused by traumatic rupture of the middle meningeal artery.

65) This disorder is believed to be caused by ammonia entering the central nervous system and disrupting neurotransmission. The patient may experience disturbances in consciousness, confusion, stupor, coma and death.

Match the descriptions with the pancreatic lesion which best describes it:

A) Mucinous cystadenoma
B) Serous microcystic adenoma
C) Pseudocyst
D) Anaplastic carcinoma
E) Pancreatoblastoma

F) Secondary adenocarcinoma
G) Ductal adenocarcinoma
H) VIPoma
I) Lymphoma
J) Pancreatic endocrine tumour

66) This well circumscribed tumour has a sponge-like appearance on macroscopic examination. On microscopic examination, it has a benign appearance and shows a clearly delineated aggregation of small cysts lined by bland cuboidal epithelium; no cytological atypia or mitotic figures are noted. It is more common in females.

67) This tumour is the commonest pancreatic tumour in childhood. It consists of solid nests of cells and acinar structures of primitive small round/polygonal cells. Squamoid corpuscles may be present.

68) This malignancy is the commonest primary pancreatic carcinoma of adults.

69) This is a fluid filled space-occupying lesion that typically occurs in the aftermath of acute pancreatitis. The lesion lacks an epithelial lining on microscopic examination.

70) Macroscopically, this tumour has cystic spaces filled with mucin-like material. Microscopically, the tumour forms cystic spaces lined by cytologically bland columnar epithelium, supported by stroma of ovarian-like morphology. No mitotic figures, cytological atypia or foci of invasion are present.

Assign each of the following descriptions the most appropriate cytological classification, from options A) – J):

A) C1   F) Thy1
B) C2   G) Thy2
C) C3   H) Thy3
D) C4   I) Thy4
E) C5   J) Thy5

71) This preparation of a breast aspirate shows hypercellularity with a purple fibrillary stroma. Numerous staghorn/antler-horn clusters of cells are present. Bare (naked) nuclei are represented.

72) This preparation of a thyroid aspirate shows hypercellularity; most of the cells are organized into microfollicles. Some of the microfollicles show nuclear crowding and haphazard cellular arrangements. Scanty colloid is noted.

73) This preparation of a breast aspirate shows several clumps of adipocytes. No atypia are seen. No other cells are present.

74) This preparation of a thyroid aspirate shows sheets of cells demonstrating nuclear crowding. Nuclear inclusions and nuclear grooves are evident in the cells. Chewing gum colloid and sparse psammoma bodies are noted.

75) This preparation from a breast aspirate is hypercellular. It shows both isolated and loosely cohesive clusters of cells of similar morphology. Pleomorphism, variable nuclear hyperchromasia and prominent nucleoli are evident.

154

The distribution and activity of lymphocytes in chronic dermatoses or lymphoproliferative disorders can be very helpful in establishing a diagnosis. Match each of the following descriptions to the disorder which best fits it:

A) Lichen nitidus
B) Lichen planus
C) Mycosis fungoides
D) Sezary syndrome
E) Psoriasis

F) Lymphocytic vasculitis
G) Erythema nodosum
H) Erythema induratum
I) Herpes folliculitis

76) A band-like junctional infiltrate of small lymphocytes; sawtooth appearance of the epidermis.

77) An epidermotropic infiltrate of small to medium sized lymphocytes, may form Pautrier's microabcesses.

78) A monomorphic reticular dermal lymphocytic infiltrate often with little junctional activity, circulating lymphocytes in peripheral blood with convoluted nuclear margins.

79) Lesions characterised by acanthosis, hyperparakeratosis, loss of the granular cell layer, and neutrophilic microabcesses. Perivascular dermal lymphocytes may be evident. Spongiform pustules of Kogol, containing lymphocytes and neutrophils, may also be present.

80) A localised lymphohistiocytic infiltrate within expanded dermal papilla, accompanied by overlying epidermal thinning. The infiltrate seems to be "grasped" by the claw-like elongation of rete ridges.

The older pathological literature includes some highly descriptive terminology for the macroscopic findings in many diseases. Match the foodstuffs used in these descriptions, options A-J, to the most appropriate pathological condition:

A) Anchovy sauce
B) Red currant jelly
C) Bread and butter
D) Sugar icing
E) Chicken fat

F) Pea soup
G) Sago
H) German sausage
I) Maple syrup
J) Café au lait

K) Nutmeg

81) Perisplenitis

82) Intussusception

83) Post mortem thrombus

84) Amoebic liver abscess

85) Neurofibromatosis

Match each of the descriptions below with one of the renal diseases, options A-J:

A) Goodpasture's syndrome
B) IgA nephropathy
C) Membranous glomerulonephritis
D) Membrano-proliferative GN*
E) Systemic lupus erythematosus

F) Focal segmental glomerulosclerosis
G) Diabetes mellitus
H) Alport's syndrome
I) Postinfectious glomerulonephritis
J) Minimal change disease/nephropathy

86) This disease is the commonest cause of nephrotic syndrome in children.

87) This renal disease is characterized by antibodies to the glomerular basement membrane and can also cause lung haemorrhage.

88) This is the commonest cause of nephrotic syndrome in adults.

89) Methenamine silver stain reveals double contours of the glomerular basement membrane in this disease. Subendothelial deposits of IgG/IgM and C3 occur in this disease. It can present as nephritic syndrome, nephrotic syndrome or as a mixed picture.

90) This disease can cause the Kimmelstiel-Wilson lesion in the glomerulus. These lesions are nodules of pink hyaline material that form adjacent to glomerular capillary loops. This is due to a marked increase in mesangial matrix from damage as a result of non-enzymatic glycosylation of proteins.

* Membrano-proliferative glomerulonephritis

Match each of the following malignancies to the environmental material most associated with it:

A) Soot
B) Azo dyes
C) Excessively hot tea
D) Vinyl chloride
E) Betel nut

F) Talc
G) Benzene
H) Arsenic
I) Hair dyes
J) Cadmium

91) Hepatocellular carcinoma

92) Lung cancer

93) Squamous cell carcinoma of the oral cavity

94) Bladder transitional cell carcinoma

95) Acute myeloid leukaemia

Assign each of the statements below to one of the adnexal skin tumours/lesions, options A-L, which is the best match:

A) Pilar sheath acanthoma
B) Syringoma
C) Myoepithelioma
D) Trichofolliculoma
E) Pilomatrixoma

F) Perifollicular fibroma
G) Eccrine cylindroma
H) Trichoepithelioma
I) Hidroacanthoma simplex
J) Hidradenoma papilliferum

K) Hair follicle naevus
L) Spiradenoma

96) This rare pilar tumour is intermediate in differentiation between a hair follicle naevus and a trichoepithelioma. Microscopy of this solitary tumour demonstrates a central follicle that radiates into numerous small follicles of varying states of maturation. A cellular stroma is present.

97) This is a poorly circumscribed dermal tumour composed of irregularly shaped islands and cords of basaloid cells ("jigsaw pattern"). The aggregations of basaloid cells show a prominent basement membrane containing type IV collagen and laminin.

98) This neoplasm is capable of causing the Borst-Jadassohn phenomenon.

99) This is a well-demarcated nodule in the dermis or lamina propria. It has a mixed papillary and cystic structure. Two types of epithelium are present – an eosinophilic tall columnar type and an underlying myoepithelial type. Decapitation secretions into the lumina may be evident.

100) This solitary dermal lesion is well circumscribed and contains basaloid cells and shadow cells.

Match each of the following descriptions, 101-105, to the vascular disorder which best matches it:

A) Buerger's thromboangiitis obliterans
B) Churg-Strauss disease
C) Kawasaki disease
D) Leukocytoclastic vasculitis
E) Takayasu's disease

F) Temporal arteritis
G) Polyarteritis nodosa
H) Wegener's granulomatosis
I) Henoch-Schonlein purpura
J) Microscopic polyangiitis

101) An allergic inflammation of small cutaneous blood vessels, characterized clinically by palpable purpura. Nuclear dust and fibrinoid necrosis may be observed.

102) A medium and small vessel autoimmune vasculitis, that involves mainly the blood vessels of the lungs, gastrointestinal system, and peripheral nerves. The individual can present with asthma as a result of the disease.

103) Recurring progressive inflammation and thrombosis of small and medium sized arteries and veins of the hands and feet, strongly associated with use of tobacco.

104) This aortitis is also known as "pulseless disease" because pulses in the upper extremities, such as the wrist pulse, may not be palpable.

105) An autoimmune disease that manifests as a systemic necrotizing medium-sized vessel vasculitis and is mainly seen in children under 5 years of age. It affects many organ systems, including the heart, vessels, skin, mucous membranes and lymph node.

Match each of the following colorectal tumours, 106-110, to the stage which best describes it:

A) Dukes Stage A   G) T1N0M0   M) T4N1M1
B) Dukes Stage B1   H) T2N1MX   N) T1N1M0
C) Dukes Stage B2   I) T2N0M0
D) Dukes Stage C1   J) TisN0M0
E) Dukes Stage C2   K) T3N0M0
F) Dukes Stage D   L) T4N2M1

106) This is a moderately differentiated adenocarcinoma of the sigmoid colon which extends into but not through the muscularis propria. The tissue shows vascular and perineural invasion. Approximately 40% of the tumour has a mucinous character. Only one node, the apical node, contains tumour metastases. The proximal, distal and serosal margins are tumour-free.
Identify the Dukes stage.

107) This rectal segment shows a polypoid area with high grade dysplasia in the mucosa without invasion. No vascular or perineural invasion is seen. The proximal, distal and circumferential margins are tumour-free. No lymph node involvement is present. No metastasis has occurred.
Identify the TNM stage.

108) Identify the Dukes classification equivalent to T3N0M0.

109) This is a poorly differentiated adenocarcinoma of the descending colon which extends through the muscularis propria and invades the visceral peritoneum. The tissue shows vascular and perineural invasion. 8 of the 10 identified lymph nodes contain tumour metastases. The proximal and distal margins are tumour-free. Liver metastases were identified radiologically and confirmed histologically.
Identify the TNM stage.

110) Assuming that a colorectal tumour is Dukes stage A, without nodal involvement or metastasis, *identify the equivalent TNM stage.*

Match each of the following descriptions, 111-115, to the testicular tumour which best matches it:

A) Seminoma
B) Yolk sac tumour
C) Teratoma
D) Choriocarcinoma
E) Adenocarcinoma

F) Spermatocytic seminoma
G) Lymphoma
H) Embryonal carcinoma
I) Small cell carcinoma
J) Leydig cell tumour
K) Sertoli cell tumour

111) This tumour has a bimodal age distribution. It is usually unilateral and can vary in size from 2-20cm. Histologically the tumour consists of three types of discohesive cells with scanty stroma. The small cells have dark staining nuclei and scant eosinophilic cytoplasm, the intermediate cells vary in size and have round nuclei with variable amounts of eosinophilic cytoplasm and the large cells have round/oval indented nuclei and spireme–like chromatin. Intratubular germ cell neoplasia is generally not present.

112) This tumour can have varying morphology and often occurs with non-seminomatous germ cell tumours. Metastatic spread is common on presentation. Cuboidal cells with clear cytoplasm and prominent nuclei are present. Mitotic activity is brisk. Characteristically, Schiller-Duval bodies are present.

113) Microscopically this tumour has a carcinomatous appearance; solid, glandular, alveolar or tubular patterns may be present. The cells may be hyperchromatic, show prominent nucleoli and frequent mitotic figures. The tumour is immunopositive for CD30 and cytokeratins. Tissue adjacent to the tumour may show intratubular embryonal carcinoma.

114) This is the commonest type of secondary malignancy in the testis.

115) This germ cell tumour has a peak age incidence of 30-40 and is notorious for being radiosensitive. It is immunopostive to PLAP and immunonegative for AFP and HCG.

Match each of the following descriptions, 115-120, to the lung disease that fits it best:

A) Lymphocytic interstitial pneumonia
B) Respiratory bronchiolitis
C) Desquamative interstitial pneumonia
D) Diffuse alveolar damage
E) Organizing pneumonia
F) Non-specific interstitial peumonia

G) Usual interstitial pneumonia
H) Farmer's lung
I) Bird fancier's lung
J) Simple coal worker's pneumoconiosis
K) Progressive massive fibrosis
L) Caplan's disease

116) This extrinsic allergic alveolitis is caused by a type III hypersensitivity reaction to warm hay containing thermophilic actinomyces.

117) This histological pattern is common in idiopathic pulmonary fibrosis. The fibrosis shows temporal heterogeneity; areas of fibroblastic foci and patchy fibrosis are present. The fibroblastic foci are characterized by numerous plump spindle cells and sparse collagen, the patchy fibrosis is characterized by hyalinised collagen of low cellularity.

118) This pattern of lung injury can occur with a rapid clinical progression after a flu-like episode. Histologically, interstitial expansion with a fibroblastic proliferation and a mixed inflammatory infiltrate are present. Hyaline membranes may be present. Most patients die from this process.

119) This industrial lung disease causes anthracosis. In its most basic form, dark nodules less than 2cm in maximum extent are present in the lung. The disease is non-neoplastic and is not a response to biological tissue. There is little decrease in lung function.

120) This histological pattern of lung disease is common in heavy smokers. It is characterized by numerous alveolar macrophages in respiratory bronchioles that extend into adjacent alveoli. The macrophages have eosinophilic cytoplasm and contain a brown finely granular pigment. There is usually an accompanying chronic inflammatory infiltrate in the bronchiole and alveolar walls.

Match each of the descriptions below, 121-125, to one of the listed causes of natural *sudden death* in adults:

A) Myocardial infarction
B) Peptic ulceration
C) Arrhythmia
D) Aortic stenosis
E) Ruptured berry aneurysm
F) Pulmonary embolus
G) Abdominal aortic aneurysm

H) Myocardial bridging
I) Haemopericardium
J) Ectopic pregnancy
K) Tension pneumothorax
L) Hypertensive cardiomyopathy
M) Mitral prolapse
N) Hypertrophic obstructive cardiomyopathy

121) At autopsy the pericardial sac is enlarged and has a dark blue tinge. Warfarin therapy and trauma are accepted causes of this pathology.

122) A 75 year old woman with remarkably good health collapses. An inexperienced junior doctor attends to her and notes an ejection systolic murmur. The patient dies shortly afterwards. She has no history of arrhythmia or ischaemic heart disease.

123) The deceased's heart was found to have a left anterior descending artery that was buried 6mm into the myocardium for 25mm of the vessel's length. No other significant pathology was found.

124) This disease is associated with polycystic kidney disease and is capable of causing sudden death. The most likely site of the critical lesion is at the junction of the anterior communicating artery and the anterior cerebral artery.

125) A 35 year old patient complains of sudden onset chest pain worsened by breathing. The Factor V Leiden test is positive. Nothing else of note is found in the past medical history.

ANSWERS

1) **B**	43) **E**	85) **J**
2) **B**	44) **B**	86) **J**
3) **E**	45) **B**	87) **A**
4) **A**	46) **E**	88) **F**
5) **B**	47) **D**	89) **D**
6) **E**	48) **B**	90) **G**
7) **D**	49) **C**	91) **D**
8) **D**	50) **D**	92) **J**
9) **C**	51) **D**	93) **E**
10) **E**	52) **A**	94) **B**
11) **D**	53) **E**	95) **G**
12) **E**	54) **K**	96) **D**
13) **E**	55) **B**	97) **G**
14) **E**	56) **C**	98) **I**
15) **D**	57) **D**	99) **J**
16) **B**	58) **E**	100) **E**
17) **A**	59) **B**	101) **D**
18) **D**	60) **A**	102) **B**
19) **D**	61) **L**	103) **A**
20) **A**	62) **B**	104) **E**
21) **C**	63) **D**	105) **C**
22) **D**	64) **H**	106) **E**
23) **D**	65) **A**	107) **J**
24) **C**	66) **B**	108) **C**
25) **C**	67) **E**	109) **L**
26) **A**	68) **G**	110) **G**
27) **B**	69) **C**	111) **F**
28) **A**	70) **A**	112) **B**
29) **D**	71) **B**	113) **H**
30) **D**	72) **H**	114) **E**
31) **B**	73) **A**	115) **A**
32) **B**	74) **J**	116) **H**
33) **D**	75) **E**	117) **G**
34) **D**	76) **B**	118) **D**
35) **C**	77) **C**	119) **J**
36) **B**	78) **D**	120) **B**
37) **C**	79) **E**	121) **I**
38) **A**	80) **A**	122) **D**
39) **C**	81) **D**	123) **H**
40) **A**	82) **B**	124) **E**
41) **E**	83) **E**	125) **F**
42) **B**	84) **A**	

Paper 4 – Answers

1) **B**

The **solitary fibrous tumour of the pleura** is CD34 positive and has a patternless architecture with a collagenous background. Solitary fibrous tumour of the pleura, previously sometimes referred to as benign mesothelioma, may occur at many body sites, but is classically associated with the pleura. There is no link with asbestos exposure and excision is curative.

2) **B**

The acronym used to remember the perinatal vertical transmission of infections from the mother to fetus is TORCH:

Toxoplasmosis
Other infections
Rubella
Cytomegalovirus
Herpes simplex virus

The *other* infections include Hepatitis B, Syphilis, Varicella-Zoster Virus, HIV, and Parvovirus B19. **Infectious mononucleosis** is not a vertically transmitted perinatal infection.

3) **E**

Farber's disease is a rare autosomal recessive lysosomal storage disease that causes an accumulation of ceramides. Affected infants show impaired mental ability and problems with swallowing. Monckeberg's medial calcific sclerosis refers to dystrophic calcification of the media of arteries, especially the radial or ulnar arteries. It may be associated with trophic foot ulceration and peripheral artery occlusive disease. Atherosclerosis is well known as a cause of arterial occlusion resulting in ischaemic damage to the myocardium, cerebrum and many other tissues. Leukocytoclastic vasculitis results from Type III immune complex mediated damage to small blood vessels, often reflecting a hypersensitivity reaction to drugs or other immunological stimuli such as autoimmune disorders. Buerger's disease, thromboangiitis obliterans, is a recurring inflammation and thrombosis of small and medium arteries and veins of the hands and feet and is strongly associated with smoking.

4) **A**

The question stem describes a biphasic tumour of epithelial cells (oncocytes) and lymphocytes; this is a **Warthin's tumour**. The appearance of this tumour under the microscope is unique. There are cystic spaces surrounded by two uniform rows of cells with centrally placed pyknotic nuclei. The cystic spaces have papillary infoldings that protrude into them. Additionally, the epithelium has lymphoid stroma with germinal centre formation.

5) **B**

Metaplasia represents a substitution of one cell type for another, resulting from a cytokine mediated "reprogramming" of tissue stem cells, rather than a change in actual mature cells. It is most commonly associated with epithelial surfaces (e.g.

metaplasia of respiratory epithelium to squamous epithelium due to the irritant effect of tobacco smoke). Barrett's or columnar lined oesophagus is an example of metaplasia that may predispose to malignant transformation as part of the increased cellular turnover. In principle metaplasia is reversible. Barrett's metaplasia can occur in connective tissue.

6) **E**
Acute pancreatitis is a cause of hypo- rather than hypercalcaemia, due to complexing of calcium with free fatty acids. The others represent the most important causes of hypercalcaemia. The mechanism of hypercalcaemia in hyperparathyroidism is excess secretion of parathyroid hormone, and may be glandular or a paraneoplastic syndrome. Destruction of cortical bone by metastatic carcinomas or myeloma may result in hypercalcaemia. Vitamin D intoxication or sarcoidosis (macrophages inside the granulomas convert vitamin D to its active form, resulting in elevated levels of 1,25-dihydroxyvitamin D).

7) **D**
Carcinoid tumours are low-grade malignancies, most of which are amenable to resection. Typical carcinoids have fewer than two mitoses per 10 high-power fields and lack necrosis whereas atypical carcinoids have between two and 10 mitoses per 10 high-power fields and do show at least focal necrosis. Five year survival rates approach 90% for typical carcinoids but are closer to 50% for atypical carcinoids. Most bronchial carcinoids do not possess secretory activity.

8) **D**
Adenocarcinoma does not in general cause paraneoplastic syndromes. Carcinoid tumours and small cell carcinoma may give rise to carcinoid syndrome. ACTH and ADH may be produced by small cell carcinoma, while squamous cell carcinoma may cause hypercalcaemia.

9) **C**
Lambert-Eaton myasthenic syndrome is an immunological disorder that results in the inhibition of voltage-gated calcium channels on the pre-synaptic membrane of the nerve-muscle (neuromuscular) junction. **50% of cases have an associated small-cell lung cancer**. Acetylcholine is prevented from being released from the presynaptic terminal that prevents the subsequent stimulation of the post-synaptic terminal, which would lead to muscle contraction. The clinical presentation often resembles myasthenia gravis.

10) **E**
Pancoast tumour refers to lung cancer in the apex of the lung, which may affect the cervical sympathetic plexus. This can result in pain (particularly in the distribution of the ulnar nerve) and Horner's syndrome (enophthalmosis, ptosis, miosis and anhydrosis).

11) **D**
Exudates are high protein secretions compared with transudates. Inflammatory diseases and cancers tend to cause exudates, whereas increased intravascular hydrostatic pressure or lower intravascular osmotic pressure tend to cause transudates.

Rheumatoid disease is an autoimmune disease that causes inflammation and an exudate.

12) E

Any tumour with involvement of the chest wall, diaphragm, mediastinal pleura, pericardium or main stem bronchus less than 2 cm from the carina is designated **T3**. Involvement of ipsilateral mediastinal or subcarinal nodes denotes **N2**. As there are no distant metastases this is **M0**.

13) E

The relationship between smoking and **bronchioloalveolar carcinoma** (BAC) is controversial; it is believed that the incidence of all primary lung cancers is increased by smoking, however BAC is believed to have the weakest association with smoking.

14) E

Borrelia burgdorferi causes Lyme disease. This is not particularly associated with immunosuppression. All of the agents are important causes of pneumonia in immunocompromised hosts.

15) D

Atypical pneumonia is commonly used to refer to an acute febrile illness with patchy inflammatory infiltrates; these pneumonias are not caused by the common bacterial or viral pathogens. Usually the inflammatory process is confined to the alveolar walls, with little exudate informed, so that there is generally a dry cough with little sputum production. The commonest causative organisms are viruses, **mycoplasma**, chlamydia and coxiella burnetti.

16) B

All the listed factors, except **Helicobacter pylori** infection, predispose to reflux oesophagitis.

17) A

Long-standing reflux oesophagitis predisposes to Barrett's oesophagus, which is a risk factor for oesophageal adenocarcinoma. All the other conditions are risk factors for squamous cell carcinoma of the oesophagus.

18) D

Turner's syndrome is a predisposing factor for pyloric stenosis. Edward's syndrome and oesophageal atresia also predispose.

19) D

Hypertrophic gastropathy refers to hypertrophy of the gastric mucosal folds. There is no association with Helicobacter infection. The condition actually represents three separate diseases: hyperplasia of the gastric glands secondary to Zollinger-Ellison syndrome, Menetrier's disease (a disorder of unknown aetiology but possibly due to mucosal over expression of growth inducing cytokines, frequently asymptomatic), and hypertrophic hypersecretory gastropathy due to hyperplasia of parietal and chief cells not due to Zollinger-Ellison syndrome, often resulting in peptic ulceration. Follicular gastritis refers to the histological pattern often produced by Helicobacter infection, with frequent lymphoid follicles.

20) **A**

Gastric marginal zone lymphoma, which accounts for approximately 5% of all gastric malignancies, is associated with trisomy 3 and t(11;18).

21) **C**

Bifidobacterium is a normal gut commensal. All the others are significant causes of bacterial enteritis. Although Mycobacterium might not seem an obvious cause, it causes a large number of cases of enterocolitis worldwide due to ingestion of contaminated milk. It causes a granulomatous inflammation of the bowel wall, which frequently results in perforation, strictures and fistulae.

22) **D**

As part of the extensive mucosal ulceration sometimes seen in the colon in ulcerative colitis, scattered islands of residual and regenerating mucosa may be seen. These may take on a polypoid appearance when seen against the flat and featureless areas of ulceration; these are referred to as **pseudopolyp**s.

23) **D**

This is a **FIGO stage III** cancer. Stage III is carcinoma that has extended into the pelvic sidewall. On rectal examination, there is no cancer-free space between the tumour and the pelvic sidewall. The tumour involves the lower third of the vagina. All cases with hydronephrosis or a non-functioning kidney are Stage III cancers.

24) **C**

Human papillomavirus, HPV, types **16**, 18, 31 and 33 are among the most important high-risk types.

25) **C**

Gynandroblastoma is a very rare tumour comprising both ovarian (granulosa and/or theca) and testicular (Sertoli and/or Leydig) cells or tissues; it is a sex cord-stromal tumour. Yolk Sac tumour, dysgerminoma and struma ovarii are germ cell tumours (the latter is an example of a monodermal teratoma). Brenner tumours are of epithelial origin.

26) **A**

In an ectopic pregnancy the embryonic implantation occurs outside the most favourable part of the uterus, most commonly in the fallopian tubes but at times also in extra-tubal locations. Rates of ectopic pregnancy have been steadily rising in recent years and the centre for disease control in Atlanta quote a rate of 2% i.e. **1 in 50** of all pregnancies.

27) **B**

Placenta accreta refers to abnormally deep implantation of the placenta into the myometrium, and carries a significant risk of antepartum haemorrhage. Placenta praevia is lower than normal implantation of the centre in the uterine canal and again carries a risk of antepartum haemorrhage. Placental abruption refers to the abnormal separation after 20 weeks of gestation and prior to birth and is the most common cause of late pregnancy bleeding. Placenta membranacea is a rare abnormality in which all or most fetal membranes remain covered by chorionic villi, because the chorion has failed to differentiate into chorion laeve and chorion frondosum. It may cause recurrent antepartum haemorrhage. Placenta profundus does not exist.

28) A

Papillary carcinoma of the thyroid is noteworthy for its good prognosis as well as for occurring in young people, frequently between 20 and 40 (with a female preponderance). Up to 50% of cases have metastases to cervical lymph nodes at the time of diagnosis, but this does not necessarily imply an adverse prognosis. The 10 year survival rate is over 95%.

29) D

Medullary carcinoma of the thyroid can occur as a component of MEN 2A or MEN 2B. None of the other thyroid tumours are part of the MEN syndromes.

30) D

Both sporadic and familial cases of **medullary carcinoma** are associated with point mutations in the RET proto-oncogene. Screening for these point mutations allows for early detection of cases of medullary carcinoma in patients with MEN 2.

31) B

10-15% of cases of primary hyperparathyroidism are caused by primary parathyroid gland hyperplasia. The figures 75-80% and <5% refer to the percentage of cases of primary hyperparathyroidism caused by parathyroid adenoma and parathyroid carcinoma, respectively.

32) B

Polycystic kidney disease is autosomal dominant; the options are autosomal recessive conditions. As a rule of thumb most inborn errors of metabolism are autosomal recessive whereas most mutations of key structural proteins are autosomal dominant.

33) D

Alkaptonuria was the first inborn error of metabolism to be described. It results from a lack of the enzyme homogentisate 1,2-dioxygenase, which results in the accumulation of homogentisic acid. This is deposited in hyaline cartilage, with resultant damage to joints, particularly the intervertebral articulations. A dark blue discolouration may be discerned in the cartilage of the patient's ears or nose.

34) D

None of the first three conditions, Down's, Edward's and Patau syndromes, carry an increased risk of these disorders. Down's syndrome is however noteworthy for the increased risk of developing acute myeloid and lymphoblastic leukaemias. **Klinefelter's syndrome**, usually associated with a 47,XXY karyotype, carries an increased risk of extragonadal germ cell tumours and autoimmune disease. Classic Turner's syndrome, 45 XO, does not appear to be associated with an increased risk of any form of malignancy.

35) C

Condylomata lata are papular lesions occurring as part of the spectrum of secondary syphilis, which is caused by the spirochaete Treponema pallidum. Verruca vulgaris, condyloma acuminatum and verruca plana are caused by human papillomaviruses. Molluscum contagiosum is a self-limiting viral skin infection, caused by a poxvirus.

36) **B**
Osteoid osteoma is a small (under 20 mm) benign bone tumour that tends to occur in the appendicular skeleton, especially within the cortex of long bones. The lesion produces prostaglandin E2 in large quantities, which causes pain that is diminished by aspirin.

37) **C**
Giant cell tumour is characterized by the presence of multinucleated osteoclast-like giant cells. In most patients, the tumours are indolent, but they may recur locally in as many as 50% of cases. Metastasis to the lungs may occur. Simple curettage is usually sufficient for treatment.

38) **A**
The lesions of **fibrous dysplasia** are benign tumours that show replacement of the medullary bone with fibrous tissue, causing the expansion and weakening of the areas of bone involved. Especially when involving the skull or facial bones, the lesions can cause externally visible deformities. Although the skull is often affected any other bone(s) can be involved. 70 to 80% of patients have lesions localized in only one bone (monostotic fibrous dysplasia), whereas others have them in many bones (polyostotic fibrous dysplasia). In 3% of cases, patients suffering from fibrous dysplasia also have endocrine diseases and skin pigmentation; the three together constitute McCune-Albright syndrome. These endocrine diseases include precocious (early) puberty, often occurring as early as 6 years of age.

39) **C**
Focal nodular hyperplasia is an entirely benign lesion whose main clinical significance is that it may be mistaken for a metastasis. It should not be confused with nodular regenerative hyperplasia, which is a reparative/regenerative process affecting the entire liver, usually seen in the context of portal hypertension. The central scar is usually clearly visible on CT or MRI, allowing accurate diagnosis without biopsy in many cases.

40) **A**
Thorotrast has been called the most carcinogenic substance known. It is a suspension containing particles of the radioactive compound thorium dioxide, ThO_2, used as a contrast medium in X-ray diagnostics in the 1930s and 40s. Thorium is an alpha particle emitter, resulting in very significant radiation exposure to patients over a long period because the half-life is 22 years. As well as hepatic angiosarcoma, exposed patients experienced a 20-fold increased risk of leukaemia.

41) **E**
As the name implies, **Aspergillus flavus** is responsible for production of aflatoxins; highly carcinogenic compounds present in contaminated groundnuts and grains. Chronic exposure significantly increases the risk of developing liver cancer.

42) B

Hepatic veno-occlusive disease is not a recognised risk factor for hepatocellular carcinoma; the other options are associated with hepatocellular carcinoma.

43) E

Gilbert syndrome is the only cause of (intra)hepatic rather than posthepatic jaundice in the listed options. It results in predominantly unconjugated hyperbilirubinaemia, whereas all the others result in predominantly conjugated hyperbilirubinaemia.

44) B

Posthepatic or obstructive jaundice is characterised by **dark urine** because of excess conjugated bilirubin in the blood and urine. **Stools are pale** because less bile is secreted into the duodenum. Alkaline phosphatase levels are raised but intrinsic liver enzymes such as aspartate transferase are not increased.

45) B

Worldwide, **chronic viral hepatitis** due to hepatitis B or C is the commonest cause of liver failure, followed by excessive alcohol intake resulting in cirrhosis.

46) E

Hepatitis C is a very rare cause of acute liver failure although a common cause of chronic hepatitis/chronic liver impairment. The other options more frequently cause acute liver failure.

47) D

Aspirin is contraindicated in children in viral illnesses because of its association with Reye's syndrome. The syndrome may also rarely occur in the absence of aspirin use.

48) B

Peliosis hepatis is a rare *pathological condition* of the liver in which there is secondary dilatation of hepatic sinusoids, usually secondary to intake of anabolic steroids or occasionally the oral contraceptive pill. Generally there are no clinical symptoms or signs on examination unless the disorder is complicated by an intra-abdominal haemorrhage. Each of the other signs listed are typical of hepatic encephalopathy.

49) C

Pancreas divisum is relatively common, occurring in up to 10% of people in some autopsy studies. It predisposes to chronic pancreatitis but is usually asymptomatic. The other conditions are rare except for ectopic pancreas, which occurs in up to 2% of people and is again usually asymptomatic. The stomach and Meckel's diverticulum are two common places for pancreatic rests.

50) D

Fusion of the kidney poles, usually the lower poles, results in **horseshoe kidney**. This is relatively common, occurring in up to 1 in 500 people. The other abnormalities are relatively rare.

51) D

Hunter syndrome is synonymous with mucopolysaccharidosis II. It is a lysosomal strorage disease that leads to progressive deterioration; it is life limiting. The pathogenesis of the disease arises from a deficiency in the enzyme L-iduronosulphate sulphatase. As a result, mucopolysaccharides (=glycosaminoglycans) are not broken down and accumulate in the cells. Early symptoms include abdominal hernias, ear infections and repeated colds. More established disease can cause mental retardation and cardiac/joint/lung function impairment.

52) A

Zellweger syndrome (cerebrohepatorenal syndrome) is a congenital leukodystrophy. Leukodystrophies cause progressive degeneration of the white matter of the brain. The pathogenesis of the disease lies in mutations of the proteins required to assemble peroxisomes. Peroxisomes are required for normal brain development and the function and formation of myelin, the whitish material that coats nerve fibers. In general, many of the major systems are affected, including the eye, the liver (enlarged), the kidney, the cartilage, the heart (malformation of the cardiovascular system) and the muscles. Craniofacial abnormalities may also occur.

53) E

The question stem describes **McArdle's disease (Glycogen storage disease type V)**. This is a glycogen storage disease that occurs because of a deficiency in the enzyme muscle phosphorylase.

54) J

The question stem describes clinical manifestations of **cystic fibrosis**. The underlying abnormality is a mutation in the gene for the cystic fibrosis transmembrane conductance regulator (CFTR). This gene is required to regulate the components of sweat, digestive juices and mucus. CFTR acts an ATP regulated chloride channel in the plasma membrane of epithelial cells. The mutation causes the production of thick mucus - as a result the sufferers are prone to inhibition of pancreatic secretions and recurrent lung infections.

55) B

The question stem describes the commonest defect that leads to **phenylketonuria**. Although occurring less commonly, a defect in the cofactor tetrahydrobiopterin can also cause phenylketonuria.

56) C

T cell prolymphocytic leukaemia, T-PPL, is a very rare disease analogous to CLL among B cell lymphomas. A CD4 positive, CD8 negative phenotype is usual.

57) D

NK/T cell lymphoma is a rare and aggressive lymphoma that is usually positive for CD8, CD56 and other cytotoxic markers such as TIA-1, granzyme B and perforin. EBV is usually readily demonstrable by ISH.

58) E

Precursor T cell acute lymphoblastic leukaemia/lymphoma, precursor T-ALL, like B-ALL, is usually TdT positive. CD10 (CALLA) and CD34 are also often expressed. CD45 is usually negative.

59) B
An IgM paraprotein is frequent in **lymphoplasmacytic lymphoma**, but may also sometimes be found in marginal zone lymphoma.

60) A
CD10 (and bcl-6) positivity, bcl-2 negativity, and a proliferative index approaching 100% are all characteristic of **Burkitt's lymphoma**.

61) L
Intracerebral haemorrhage is associated with systemic hypertension in more than 50% of cases. Charcot-Bouchard microaneurysms are dilations of small vessels less than 0.3 mm in maximum extent; these are caused by hypertension. "Intraparenchymal "s a phrase often used synonymously with intracerebral haemorrhage (the former assumes no ventricular bleeding).

62) B
Central pontine myelinolysis, CPM, is frequently seen in patients with alcoholism, but may also occur in patients with severe liver disease, severe burns, malnutrition, electrolyte disorders, and in hyperemesis gravidarum. It is believed to occur as a result of osmotic damage to the myelin sheath of the nerve cells in the brainstem.

63) D
The implication of the question stem is that the individual is trying to scratch a sun damaged area of skin. New onset of seizures in adult life mandates an MRI search to look for mass lesions, of which **metastatic carcinoma** is common. In this case metastatic melanoma would be considered. Secondary brain tumours can cause a range of symptoms:

- Headaches
- Weakness
- Changes in behaviour
- Seizures
- Symptoms of raised intracranial pressure
- Nausea
- Confusion
- Malaise

64) H
Extradural/epidural haemorrhage is usually caused by trauma to the middle meningeal artery. The middle meningeal artery runs **in** the dura and traumatic stress lines can rupture the artery. The bleeding occurs into the potential epidural space under arterial blood pressure. Radiological investigation will show the expanding haematoma to have a smooth outline. Clinically, the patient may be lucid after the original trauma but can deteriorate rapidly. Hence an epidural hematoma is a surgical emergency.

65) A
The question stem is describing **hepatic encephalopathy**. The hepatic impairment means that nitrogenous waste is not appropriately metabolized, leading to the accumulation of ammonia in the blood that can cross the blood-brain barrier.

66) **B**

A **serous microcystic adenoma** is described in the question stem. This rare benign lesion occurs mainly in young to middle aged females. The prognosis is good, partly because malignant transformation is rare.

67) **E**

Pancreatoblastoma is described in the question stem. 50% of cases occur in the head of the pancreas. This is an exceptionally rare "small round blue cell tumour" of childhood. Prognosis can be poor, particularly if metastasis has occurred.

68) **G**

Ductal adenocarcinoma is the commonest primary pancreatic carcinoma of adults. 60-70% occur in the head of the pancreas. It is a malignancy that tends to occur in inviduals 60-80 years of age. Smoking is a known risk factor.

69) **C**

A **pseudocyst** is described in the question stem. It occurs as a complication of pancreatitis and can often be identified radiologically. Management may require surgical drainage or excision. A pseudocyst has no epithelial lining whereas a conventional cyst has an epithelial lining.

70) **A**

A mucinous cystic neoplasm that has no atypia is described in the question; this amounts to a **mucinous cystadenoma**. Mucinous cystadenoma shows large multilocular or, in rare cases, unilocular cysts lined by tall, mucin-producing cells, which may form papillae. Thorough sampling is needed to avoid missing a cystadenocarcinoma.

71) **B**

The cytology report is consistent with a **fibroadenoma – a benign** lesion. Hence the most appropriate classification is **C2**.

C1 – Inadequate
C2 – Benign
C3 – Equivocal
C4 – Suspicious of malignancy
C5 – Malignancy

72) **H**

The cytology report shows suspicious appearances because of the hypercellularity, the numerous microfollicles and the focal atypia. The appearances are suspicious of a follicular neoplasm (neoplasia cannot be excluded). The correct classification is **Thy3**.

Thy1 – Non-diagnostic/inadequate.
Thy2 – Non-neoplastic.
Thy3 – All follicular lesions and aspirates in which neoplasia cannot be excluded.
Thy4 – Abnormal and suspicious of malignancy.
Thy5 – Diagnostic of malignancy.

73) A
The cytology report indicates only normal adipocytes. This is an **inadequate** sample.

74) J
The cytology report indicates most of the cytological features of papillary carcinoma of the thyroid; a **Thy5** classification.

75) E
The cytology report is consistent with a **ductal carcinoma** of the breast (the hypercellularity, cytological atypia and discohesive nature of the cell clusters are in favour of the diagnosis). The correct classification is **C5**.

76) B
The description refers to the "lichenoid" reaction pattern, which may be seen in a wide variety of dermatoses ranging from lichenoid drug reactions to **lichen planus** and lichenoid graft-versus-host disease.

77) C
Epidermotropism (lymphocytes invading the epidermis) is a characteristic feature of **mycosis fungoides**, particularly when it is not associated with inflammatory epidermal changes.

78) D
The infiltrate in **Sezary syndrome** is often deeper than in MF and frequently there is no epidermotropism, in contrast to MF.

79) E
All these features are characteristic of **psoriasis**, although psoriasiform reactions may also show some of the described features.

80) A
Lichen nitidus is a chronic inflammatory dermatosis of unknown aetiology, presenting with 1-2 mm, pink or brown papular lesions, sometimes present in linear arrangements. They are usually asymptomatic and self-limiting, but may be pruritic.

81) D
Sugar-icing spleen refers to splenic capsular thickening (producing a glazed white appearance) due to perisplenitis.

82) B
The localised ischaemia produced by intussusception may produce haemorrhagic stools, resembling **redcurrant jelly**.

83) E
Post mortem thrombus separates into red blood cells and fibrin, the latter resembling **chicken fat**.

84) A
The brown semi-liquid material in amoebic liver abcesses has been compared to **anchovy sauce**.

85) **J**

White coffee coloured (**café au lait**) macules are characteristic of neurofibromatosis.

86) **J**

Minimal change disease is the commonest cause of nephrotic syndrome in children; under light microscopic a relevant biopsy may have normal renal apprearances, however, under electron microscopy podocyte effacement/fusion can be identified. The latter is characteristic of the disease.

87) **A**

The question stem describes **Goodpasture's syndrome**. It is an autoimmune disease responding to the antigenic α3 chain of collagen IV. In contrast an *inherited defect* in the same protein can lead to Alport syndrome.

88) **F**

Focal segmental glomerulosclerosis (also termed focal glomerular sclerosis, focal nodular glomerulosclerosis) is the commonest cause of nephrotic syndrome in adults. A renal biopsy is required for definitive diagnosis.

Key Causes of Nephrotic Syndrome:
Minimal change nephropathy/disease
Membranous glomerulonephritis
Membranoproliferative glomerulonephritis
Focal segmental glomerulosclerosis
Diabetes Mellitus
Amyloidosis
Neoplasia – lymphoma/carcinoma
Endocarditis
Ployarteritis nodosa
Systemic lupus erythematosus
Sickle cell anaemia
Malaria
Drugs (e.g. penicillamine or gold).

89) **D**

Membranoproliferative glomerulonephritis (MPGN) can reveal double contouring of glomerular basement membrane and subendothelial immunological deposits (IgG/IgM and C3) as indicated in the question. MPGN type II shows the dense deposit disease of C3.

90) **G**

The question stem describes the Kimmelstiel-Wilson lesion of longstanding **diabetes mellitus**. Nodular glomerulosclerosis is an alternative term for the lesion. The pink hyaline material is PAS positive.

91) **D**

Vinyl chloride used in PVC manufacture is a known cause of primary liver cancer.

92) **J**

Exposure to **cadmium** during industrial manufacturing is a rare risk factor for primary lung cancer.

93) **E**
Saffrole, a compound in **betel nuts**, is a cause of oral squamous cell carcinoma.

94) **B**
Azo dyes, most since banned, are a known carcinogen for transitional cell carcinoma, TCC. Azo dyes can be metabolically broken down to aniline a potent TCC carcinogen.

95) **G**
Benzene in petrol, solvents and other fluids is a recognised cause of **acute myeloid leukaemia**, AML.

96) **D**
This skin adnexal tumour is a **trichofolliculoma**. The central hair follicle and cellular stroma facilitate the distinction from a basal cell carcinoma.

97) **G**
This skin adnexal tumour is an eccrine **cylindroma**. It is a solitary head and neck tumour that shows a female preponderance.

98) **I**
The Borst-Jadassohn phenomenon refers to clonal nests of tumour cells in the epidermis. The classical list of skin lesions that can cause this effect includes:

Intraepidermal carcinoma (IEC)
Melanoma
Seborrheic keratosis
Eccrine poroma/porocarcinoma
Paget's disease of the breast and extramammary Paget's disease
Hydroacanthoma simplex
Sebaceous carcinoma

99) **J**
The question stem describes **hidradenoma papilliferum**. Hidradenoma papilliferum is a benign apocrine lesion usually found in the vulval/perianal areas.

100) **E**
The described lesion is a **pilomatrixoma**; "calcifying epithelioma of Malherbe." This is a benign follicular neoplasm.

101) **D**
Leukocytoclastic or hypersensitivity **vasculitis** is a small-vessel disease with many causes. Drug reactions is are important causes. Neoplasia, infections and ANCA positive vasculides can also cause leukocytoclastic vasculitis. No cause is identified in up to 50% of patients.

102) **B**

Together with Wegener's granulomatosis and microscopic polyangiitis, this is one of three important necrotizing systemic small-to-medium sized vessel vasculitides associated with antineutrophil cytoplasm antibodies (ANCAs). Eosinophilia and asthma are part of the disease manifestation in **Churg-Strauss disease**.

103) A
Buerger's disease, thromboangiitis obliterans, is a recurring progressive vasculitis and thrombosis of small and medium sized arteries and veins of the hands and feet, strongly associated with smoking.

104) E
Takayasu's arteritis is a large vessel granulomatous vasculitis with massive intimal fibrosis and vascular narrowing. Patients are often young or middle-aged women of Asian descent (China/Japan/Korea). It mainly affects the aorta and pulmonary arteries.

105) C
Intravenous immunoglobulin is the standard treatment for **Kawasaki disease** and is administered in high doses with marked improvement usually noted within 24 hours. This is one of the few conditions where aspirin treatment in indicated in children.

106) E
Dukes stage C2 is defined as a primary colorectal adenocarcinoma that extends into the muscularis propria but does not penetrate through it, together with tumour metastasis to the apical lymph node. In the case of this tumour, this is equivalent to TNM stage T2N1MX.

Dukes A:	Limited to mucosa
Dukes B1:	Extending into muscularis propria but not penetrating through it; nodes not involved.
Dukes B2:	Penetrating through muscularis propria; nodes not involved.
Dukes C1:	Extending into muscularis propria but not penetrating through it. Nodes involved but apical node tumour-free.
Dukes C2:	Penetrating through muscularis propria. Node(s) involved, including the apical node.
Dukes D:	Distant metastatic spread

107) J
The question stem describes a carcinoma in situ. No lymph node involvement or distant metastasis is present. This amounts to **TisN0M0**.

108) C
The tumour has gone through the muscularis propria but has not involved lymph nodes or caused distant metastasis. This is equivalent to **Dukes stage B2**.

109) L
The peritoneal penetration raises the stage of the tumour to T4, because there are more than four regional nodes involved, N=2. The liver metastasis has been confirmed histologically. Hence this amounts to **T4N2M1**.

110) G
A simple Dukes A tumour is equivalent to **T1N0M0**.

111) F
The question stem describes a **spermatocytic seminoma**. The bimodal distribution refers to affected individuals being in their 20s or over 50. The differential diagnosis includes classical seminoma and lymphoma. The classical seminoma usually excites a lymphocytic/granulomatous response and usually has an ITGCN (intratubular germ cell neoplasia) component. Lymphoma usually has an interstitial growth pattern and an appropriate immunophenotype.

112) B
This is a **yolk sac tumour**. Schiller-Duval bodies are helpful diagnostically, if present. Immunopositivity with AFP is also useful diagnostically; however the reaction can be weak or absent.

113) H
This is the description of **embryonal carcinoma** (malignant teratoma undifferentiated). This poorly differentiated epithelial malignancy usually occurs in individuals under 30.

114) E
Adenocarcinomas, of all origins, are the commonest type of secondary tumour in the testis. Prostate, kidney and lung are the most likely primary sites.

115) A
Classical **seminoma** is a malignant germ cell tumour of the testis that is very radiosensitive.

116) H
The question stem describes **Farmer's lung**. This is a reaction to inhaled allergenic organic material; specifically a response to actinomyces proteins.

117) G
This is the description of **usual interstitial pneumonia**. It is the commonest histological pattern seen in idiopathic pulmonary fibrosis. Synonyms for idiopathic pulmonary fibrosis include:

Cryptogenic fibrosing alveolitis
Diffuse fibrosing alveolitis
Usual interstitial pneumonitis
Interstitial diffuse pulmonary fibrosis
Fibrosing alveolitis

118) D
This is **diffuse alveolar damage**. This histological pattern is seen in acute respiratory distress syndrome, which is synonymous with acute interstitial pneumonia.

119) J

Simple coal worker's pneumoconiosis contains macules of coal that are less than 2mm and nodules that are less than 2cm in maximum extent. The nodules contain collagen and coal. There is little decrease in lung function. In comparison, progressive massive fibrosis demonstrates fibrotic nodules of coal that are 2-10cm in diameter and have a necrotic centre. There is a 10% risk of disease progression to chronic lung disease, pulmonary hypertension and cor pulmonale.

120) **B**
The question stem describes **respiratory bronchiolitis**. The disease is on a spectrum that overlaps with desquamative interstitial pneumonia; both cause macrophage accumulation and are associated with smoking.

121) **I**
Haemopericardium refers to the accumulation of blood in the pericardial space. If it continues then the clinical result of cardiac tamponade occurs, as the external pressure compromises effective contraction of the myocardium. An experienced pathologist can identify haemopericardium from the macroscopic appearance of the pericardium.

122) **D**
Considering the age of the patient, the lack of significant medical history and the apparent presence of an ejection systolic murmur, the most likely diagnosis is critical **aortic stenosis**. The calcification that occurs with time accumulates in this age group to cause impairment of aortic outflow pressure. Such individuals are prone to sudden death.

123) **H**
For **myocardial briging** to be pathologically significant the coronary artery must be buried in the myocardium at a depth of at least 5mm for at least 20mm of its length. Less than this, and the effect of compression of the coronary artery with cardiac contraction, is unlikely to have a significant functional effect.

124) **E**
The question stem is describing a **ruptured berry aneurysm** that can cause death through subarachnoid haemorrhage and is associated with adult polycystic kidney disease. Approximately 40% of these aneurysms are situated between the anterior communicating artery and the anterior cerebral artery.

125) **F**
Factor V is a clotting factor that has a normal physiological role. However mutation can cause a variant termed Factor V Leiden that is difficult for the body to inactivate (it resists degradation by Protein C). Such mutations are associated with diseases of hypercoagulability such as **pulmonary embolism**. Pulmonary embolism causes a pleuritic chest pain; pain that is worsened by breathing.

PAPER 5

1) The lowest international incidence rates for gallbladder cancer are in:

(A) Peru
(B) India
(C) Britain
(D) Chile
(E) Native American populations in Alaska.

2) Graft versus host disease is an example of which type of hypersensitivity reaction?

(A) Type I
(B) Type II
(C) Type III
(D) Type IV
(E) Type V

3) Subacute bacterial endocarditis is an example of which type of hypersensitivity reaction?
(A) Type I
(B) Type II
(C) Type III
(D) Type IV
(E) Type V

4) The following are all causes of acute pancreatitis, except:

(A) Scorpion venom
(B) Hyperlipidaemia
(C) Hypocalcaemia
(D) Gallstones
(E) Alcohol

5) In macronodular cirrhosis, most of the nodules are larger than:

(A) 2mm
(B) 3mm
(C) 4mm
(D) 5mm
(E) 10mm

6) Which of the following is not caused by an RNA virus?

(A) Hepatitis A
(B) Hepatitis B
(C) Hepatitis C
(D) Hepatitis D
(E) Hepatitis E

7) Slit-lamp examination may reveal Keyser-Fleischer corneal rings in patients with which of the following conditions?

(A) Haemochromatosis
(B) Wilson's disease
(C) Alcoholic hepatitis
(D) Reye's syndrome
(E) Galactosaemia

8) The following carcinomas all preferentially invade veins rather than lymphatics, except:

(A) Follicular carcinoma of the thyroid
(B) Hepatocellular carcinoma
(C) Renal cell carcinoma
(D) Pancreatic carcinoma
(E) Adrenal cortical carcinoma

9) Which of the following causes preferential toxicity in the periportal zone of the liver?

(A) Paracetamol overdose
(B) Phosphorus poisoning
(C) Ischaemia
(D) Yellow fever
(E) Carbon tetrachloride toxicity

10) Which of the following inflammatory cells is frequently increased in the blood during helminthic infections?

(A) Eosinophils
(B) Lymphocytes
(C) Basophils
(D) Neutrophils
(E) Plasma cells

11) A "keloid" results from which of the following pathological processes?

(A) Neoplasia
(B) Infarction
(C) Cellular hyperplasia
(D) Scarring
(E) Cellular metaplasia

12) The t(8:14) translocation of Burkitt's lymphoma places which of the following genes under the control of the IgG promoter?

(A) c-sis
(B) myc
(C) p16

(D) ras
(E) src

13) Histological examination of an artery showing temporal arteritis reveals granulomas, often including giant cells, attacking the:

(A) Adventitia
(B) Myointimal cells
(C) Internal elastic membrane
(D) Endothelial cells
(E) Smooth muscle

14) Neutrophil rich granulomas or stellate microabscesses are characteristic of the following bacterial infections of lymph nodes, except:

(A) Tuberculosis
(B) Yersinia
(C) Brucellosis
(D) Lymphogranuloma venereum
(E) Tularemia

15) Which of the following statements is true of lymphoblastic lymphoma?

(A) It may be distinguished immunophenotypically from lymphoblastic leukaemia.
(B) The majority of cases of lymphoblastic lymphoma are of B-cell origin.
(C) CD 45 is typically positive within the lymphoma cells by immunohistochemistry.
(D) If more than 50% lymphoid blasts are present in the bone marrow, this is diagnostic of the leukaemic form rather than lymphoma.
(E) TdT is typically positive within the lymphoma cells by immunohistochemistry.

16) Recurrent angioedema with severe swelling of the soft tissues of the airway, may be due to hereditary deficiency of:

(A) Acetylcholinesterase
(B) Protein C
(C) Mast-cell deconvertase
(D) C1 esterase inhibitor
(E) Alpha-1 protease inhibitor

17) Pernicious anaemia is most likely to cause lesions in which part of the central nervous system?

(A) Central gray matter of the spinal cord
(B) Hippocampus
(C) Cerebellum
(D) Posterior columns of the spinal cord
(E) Substantia nigra

18) Which of the following carcinomas is most prone to spread via vascular invasion?

(A) Papillary thyroid carcinoma
(B) Pancreatic carcinoma
(C) Follicular thyroid carcinoma
(D) Adrenal cortical carcinoma
(E) Parathyroid carcinoma

19) Which of the following malignancies is most likely to give rise to generalised itching in the sufferer during a hot shower?

(A) Small cell carcinoma of lung
(B) Squamous cell carcinoma of lung
(C) Essential thrombocythaemia
(D) Polycythaemia rubra vera
(E) Melanoma metastatic to the liver

20) Which of the following tumours is most likely to be associated with chest pain on consumption of alcohol?

(A) Thymoma
(B) Thymic carcinoma
(C) Hodgkin's lymphoma
(D) non-Hodgkin's lymphoma
(E) Neurofibroma

21) Which of the following tumours is commonly associated with red cell aplasia?

(A) Insulinoma
(B) Glucagonoma
(C) Lymphoblastic leukaemia
(D) Thymoma
(E) Medulloblastoma

22) Severe ischaemia of the brain is associated with which type of necrosis?

(A) Caseous
(B) Liquefactive
(C) Coagulative
(D) Fibrinoid
(E) Fat necrosis

23) A high fibre diet is believed to be protective against the following pathological conditions, except:

(A) Sigmoid volvulus
(B) Colonic carcinoma
(C) Varicose veins
(D) Diverticulosis
(E) Rectal carcinoma

24) Which of the following causes jaundice that is characterised by elevated levels of unconjugated bilirubin?

(A) Rotor syndrome
(B) Dubin-Johnson syndrome
(C) Exogenous oestrogens
(D) Gilbert's syndrome
(E) Pancreatic carcinoma

25) Coagulopathy in liver failure results from failure to synthesise the following clotting factors, except:

(A) II
(B) V
(C) VII
(D) VIII
(E) X

26) Hepatic encephalopathy is associated with the presence of which of the following within the brain?

(A) Negri bodies
(B) Alzheimer type II glia
(C) Pick bodies
(D) Microglial proliferation
(E) Auer rods

27) Familial risk factors for pancreatic carcinoma include all of the following except:

(A) Familial chronic pancreatitis
(B) Peutz-Jegher syndrome
(C) Lynch/HNPCC syndrome
(D) Fabry's syndrome
(E) BRCA2 gene

28) The following features are common histological features of diabetic nephropathy, except:

(A) Capsular drop
(B) Kimmelstiel-Wilson lesion
(C) GBM thickening
(D) Fibrin cap
(E) Membranoproliferative glomerulonephritis

29) Which of the following histological features is said to be characteristic of mantle cell lymphoma?

(A) Follicular colonisation
(B) Monocytoid differentiation
(C) Plasmacytoid differentiation

(D) Presence of eosinophilic histiocytes
(E) Presence of reactive follicles

30) Which of the following conditions results in dense fibrous, often calcified, plaques within the penile shaft?

(A) Peyronie's disease
(B) Hypospadias
(C) Epispadias
(D) Phimosis
(E) Paraphimosis

31) Sister Mary Joseph's nodule refers to cancer metastatic to the

(A) Umbilicus
(B) Vertex of the scalp
(C) Left supraclavicular lymph node
(D) Nipple
(E) Peritoneal surface

32) Which of the following is not an indication for a lumbar puncture?

(A) Suspected viral meningitis
(B) Suspected multiple sclerosis
(C) Suspected ruptured berry aneurysm
(D) Suspected bacterial meningitis
(E) Suspected primary CNS lymphoma

33) Leishmaniasis is transmitted by the bite of which arthropod vector?

(A) Mosquito
(B) Sand fly
(C) Bed bug
(D) Rat flea
(E) Tsetse fly

34) American trypanosomiasis (Chagas' disease) is transmitted by the bite of which arthropod vector?

(A) Tick
(B) Rat flea
(C) Chigger
(D) Reduviid bug
(E) Tsetse fly

35) Which of the following is NOT a feature of tetralogy of Fallot?

(A) Right ventricular hypertrophy
(B) Ventricular septal defect
(C) Left ventricular hypertrophy

(D) Pulmonary stenosis
(E) Overriding aorta

36) Which of the following skin tumours is not usually painful?

(A) Eccrine spiradenoma
(B) Glomus tumour
(C) Campbell de Morgan Spot
(D) Schwannoma
(E) Angiolipoma

37) Considering intestinal polyps, the presence of a branching core of smooth muscle within the lamina propria is typical of a:

(A) Villous adenoma
(B) Tubulovillous adenoma
(C) Peutz-Jegher polyp
(D) Juvenile polyp
(E) Pseudopolyp of ulcerative colitis

38) All the following are poor prognostic factors for gastrointestinal stromal tumours except:

(A) The presence of a myxoid stroma
(B) The presence of stromal hyalinization
(C) High cellularity
(D) Mitotic count over 5 per 50 high-power fields
(E) Invasion of the overlying mucosa

39) The following statements are true of acoustic neuroma, except:

(A) It is histologically identical to neurofibroma.
(B) It accounts for 5 to 10% of all intracranial neoplasia.
(C) It occurs in neurofibromatosis type I.
(D) It occurs in neurofibromatosis type II.
(E) It may be associated with increased intracranial pressure.

40) The CPD scheme of the Royal College of Pathologists mandates the accumulation of how many points over each five-year period?

(A) 100
(B) 500
(C) 250
(D) 125
(E) 5000

41) The body that accredits UK pathology laboratories, Clinical Pathology Accreditation (CPA), is co-owned by all the following stakeholder organisations, except:

(A) Royal College of Pathologists
(B) Institute of Healthcare Management
(C) Institute of Biomedical Science
(D) Association of Clinical Pathologists
(E) National Institute for Clinical Excellence

42) Myelolipomas are rare, benign mesenchymal tumors composed of mature adipose tissue and hematopoietic cells in varying proportions. They occur most commonly in which of the following organs:

(A) Lung
(B) Bone
(C) Kidney
(D) Liver
(E) Adrenal

43) Which of the following is not an acceptable step for a pathologist to take in relation to tissue removed at coroner's autopsy once the coroner's authority has ended?

(A) Return the tissue to the family.
(B) Dispose of the material.
(C) Obtain the family's consent to retain the tissue for research.
(D) Retain the tissue in storage unless requested not to, in case of medical legal issues.
(E) Obtain the family's consent to retain the tissue for education and training.

44) Malignant melanoma is frequently positive with the following immunohistochemical markers, except:

(A) CD117
(B) HMB45
(C) CD99
(D) S100
(E) CD25

45) "Bednar tumour" refers to the following lesion:

(A) Pigmented dermatofibrosarcoma protuberans (DFSP)
(B) Giant cell fibrosarcoma
(C) Atypical lipoma of the back
(D) Haemangiopericytoma of the chest wall
(E) Pigmented synovial sarcoma

46) A tumour composed partly or mainly of coarsely multivacuolated fat cells with small, central nuclei and no atypia is likely to be a:

(A) Myelolipoma
(B) Hibernoma
(C) Angiolipoma

(D) Atypical lipoma
(E) Liposarcoma

47) The following circumstances surrounding a death are indications for a coroner's inquest, except:

(A) Death relating to an occupational disease.
(B) Death where the deceased has not been seen by a doctor within the past three months.
(C) Death in police custody.
(D) Death in childbirth.
(E) Death occurring while undergoing an operation or under anaesthetic.

48) Positivity for the following markers is characteristic of adrenal cortical carcinoma, except:

(A) Calretinin
(B) Inhibin
(C) Melan-A
(D) Vimentin
(E) CD117

49) The description "symplastic" in relation to tumours (usually uterine leiomyoma), means:

(A) Marked pleomorphism but rare mitotic figures.
(B) Showing invasion of a tumour capsule but not penetration.
(C) Showing vascular invasion.
(D) Showing numerous mitotic figures but no cellular atypia.
(E) Showing necrosis owing to the tumour outgrowing its blood supply.

50) Approximately 60% of osteosarcomas occur in the following locations:

(A) Axial skeleton.
(B) Either side of the knee joint.
(C) Either side of the hip joint.
(D) The mandible.
(E) The pelvis, sacrum and lumbar vertebrae.

Match the descriptions below to the most appropriate options, A-K:

A) Dysplasia   F) Apoptosis   K) Reperfusion
B) Hyperplasia   G) Infarction
C) Hypoplasia   H) Hypertrophy
D) Metaplasia   I) Atrophy
E) Necrosis   J) Aplasia

51) Programmed cell death.

52) This is what happens to the calf muscles after a broken leg is placed in a cast.

53) Disordered growth in cells, tissues or organs.

54) The reason why a thrombolytic or angioplasty is given shortly after the onset of an acute myocardial infarction.

55) Congenital absence of an organ.

Match the descriptions below to the most appropriate options, A-K:

A) G_1 F) Permanent   K) G_o
B) G_2 G) Labile   L) Dyclin
C) S   H) Cyclin
D) M   I) Mitosis
E) Quiescent   J) Meiosis

56) The process by which a diploid mother cell is converted into two haploid daughter cells.

57) During this phase of the cell cycle most of the DNA synthesis occurs.

58) These differentiated cells have no significant rate of turnover in adult life; neurons and skeletal muscle cells are examples of such cells.

59) This cofactor allows kinases to phosphorylate retinoblastoma protein to control the cell cycle.

60) Quiescent cells stay at this point in the cell cycle.

A) Crohn's disease
B) Ulcerative colitis
C) Diverticulosis
D) Diverticulitis

E) Perianal abscess
F) Hyperplastic polyp
G) Adenomatous polyp
H) Inflammatory pseudopolyp

I) Colorectal carcinoma
J) Colorectal adenoma
K) Familial adenomatous
       polyposis

Each of the descriptions below refers to the appearance of a large bowel specimen. Match the description to the most likely diagnosis.

61) A rectal resection from a 50 year old male smoker demonstrates a single polypoid ulcerated and obstructing mass, 9cm in maximum extent. The large intestine proximal to the mass is dilated. The mesorectum contains several enlarged and palpable lymph nodes.

62) This section of colon from a 60 year old woman shows multiple outpouchings of the wall of the large bowel. All of the outpouchings are intact except one, which appears to be perforated and demonstrates an adjacent purulent exudate.

63) A colorectal specimen from a 12 year old boy shows numerous polypoid outgrowths of the mucosa, numbering at least 250. Each polyp is less than 1.5cm in size.

64) A 45 year old woman presents with a history of sporadic bouts of bloody diarrhoea. Her large intestinal resection shows reddened mucosa that is focally ulcerated. These areas are scattered discontinuously from the caecum to the rectum. No perforations are noted. Microscopy reveals small well formed granulomata within areas of inflammation.

65) A patient with a longstanding history of bloody diarrhoea yields a resection specimen that shows fat wrapping and, on opening, mucosal cobblestoning.

A) 24q12
B) 23q11
C) 22q11
D) Glucocerebrosidase
   deficiency

E) HLA B27
F) HLA DR3
G) HLA DR4
H) IgA

I) BRCA-2
J) MSH6
K) TERT
L) α-1,4-glucosidase
   deficiency

Which of the above molecular features is best matched with each of the descriptions below?

66) This disease had an onset in childhood and was associated with a significant enlargement of the heart and liver and a generalized loss of muscle tone. Muscle biopsy revealed a lysosomal defect.

67) The HLA association most linked with ankylosing spondylitis.

68) DiGeorge syndrome.

69) The HLA association most linked with rheumatoid arthritis.

70) A tumour suppressor gene.

A) I	F) VI	K) XI
B) II	G) VII	L) XII
C) III	H) VIII	M) XIII
D) IV	I) IX	N) XIV
E) V	J) X	O) XV

Match the type of collagen below to the descriptions above:

71) This is the commonest type of collagen in basement membranes.

72) The collagen defect in classical Ehlers-Danlos syndrome.

73) This type of collagen is the target antigen in Goodpasture's syndrome.

74) This brittle bone disease has defective collagen of this type.

75) This is the commonest collagen in cartilage.

A) Liver
B) Skeletal muscle
C) Thyroid
D) Ovary

E) Melanocytes
F) Endothelium
G) Pancreas
H) TTF1

I) Kidney
J) Prostate
K) Lung

Each of the immunohistochemistry results below refers to a malignancy of unknown origin. What is the most likely primary site of each malignancy?

76) An adenocarcinoma that is CA19.9 positive, CK19 positive, CK7 positive and CK20 positive.

77) An adenocarcinoma that is RCC positive, alphabetacrystallin positive, CK7 negative, CK20 positive and CD10 positive.

78) An adenocarcinoma that is CK7 positive, CK20 negative, CEA positive and TTF1 positive.

79) This spindle cell malignancy is Myoglobin, MyoD1 and Desmin positive.

80) This adenocarcinoma is PSA and PSAP positive.

A) Mitral stenosis   F) Tricuspid regurgitation   K) Mitral valve
B) Mitral regurgitation   G) Pulmonary stenosis   L) Aortic valve
C) Aortic stenosis   H) Pulmonary regurgitation M) Tricuspid valve
D) Aortic regurgitation   I) Synthetic cardiac valve   N) Pulmonary valve
E) Tricuspid stenosis   J) Bioprosthetic cardiac valve

Select the cardiac valve/disease that most closely matches the descriptions below:

81) A cardiac valve disease that is a cause of sudden death in adults and is associated with a pansystolic murmur.

82) Intravenous drug users are most likely to acquire infective endocarditis at this valve.

83) Corrigan's sign and De Musset's sign are features of this cardiac valve disease.

84) Congenital parachute valve can cause this valve disease.

85) Primary valvular disease cause of sudden death in an adult that has an ejection systolic murmur.

A) Aniline dye workers
B) Helicobacter pylori
C) Water
D) Vitamin D
E) Hypercholesterolaemia

F) Temperature
G) High fibre diet
H) Barrett's oesophagus
I) Folic acid
J) House dust mite

K) Oxygen
L) Head trauma
M) Asbestos
N) Ultraviolet radiation
O) Antiphospholipid
       antibodies

Select the strongest risk factors for the diseases listed:

86) Oesophageal adenocarcinoma

87) Repeated miscarriages

88) Transitional cell carcinoma of the bladder

89) Duodenal ulcer

90) Primary lung adenocarcinoma

A) TSH
B) TRH
C) T4
D) T3
E) Itch
F) Sweating
G) Psychosis
H) Tremor

I) Thyrotoxicosis
J) Goitre
K) Irritability
L) Diarrhoea
M) T3 Receptor
N) T4 receptor
O) TRH receptor
P) TSH receptor

Q) Diffuse toxic hyperplasia
R) Toxic multinodular goitre
S) Toxic adenoma
T) Hyperthyroidism
U) Exophthalmos
V) Infiltrative Dermopathy

Match the descriptions below with the best answer from the list above.

91) Caused by clinically symptomatic hyperthyroidism.

92) Graves disease classically has antibodies to this.

93) Commonest cause of thyrotoxicosis.

94) The clinical triad used in the diagnosis of Graves' disease includes this synonym for pretibial myxoedema.

95) This is the result of retro-orbital inflammation and oedema.

A) Osteoarthritis   F) Ganglion   K) Osteoporosis
B) Osteomyelitis   G) Pseudogout   L) Osteogenesis imperfecta
C) Septic arthritis   H) Metastasis
D) Osteomyelitis   I) Ankylosing spondylitis
E) Gout   J) Rheumatoid arthritis

Match each of the descriptions below with one of the diseases listed above.

96) This seronegative disease can cause "bamboo spine" and is associated with mitral regurgitation and aortic regurgitation.

97) This disease can cause acute joint pain and is associated with the deposition of negatively birefringent crystalline material.

98) Associated with HLA-DRB1, proximal interphalangeal joint pain, boutonniere and swan neck deformities.

99) This disease is more common in women and predisposes to vertebral spine fractures and distal forearm fractures.

100) A 16 year old boy complained of recent onset knee pain; the aspirate from the knee revealed abundant neutrophils.

A) Giant cell arteritis
B) Takayasu
C) Kawasaki's disease
D) Buerger's disease
E) Polymyositis

F) Systemic lupus erythematosus
G) Henoch-Schonlein purpura
H) Wegener's granulomatosis
I) Churg-Strauss syndrome
J) Polyarteritis nodosa

Match each of the scenarios below with the most likely diagnoses from the list above.

101) An adult patient has new severe asthma attacks; a lung biopsy reveals vasculitis and eosinophilia. Blood tests show pANCA positivity.

102) A 4 year old boy presents with a recent history of chest pain that is eventually shown to be cardiac in origin.

103) A patient with chronic sinusitis is found to have deteriorating lung and renal function. During an exacerbation of the disease, blood tests showed cANCA positivity. A lung biopsy confirmed small vessel vasculitis.

104) A 65 year old man presents to his GP with recent onset of severe headaches that are worsened when he combs his hair. He sufferers from polymyalgia rheumatica.

105) A 30 year old male is a heavy smoker. Biopsy of his radial and tibial arteries confirms a medium vessel vasculitis.

A) Amyloidosis
B) Diabetes mellitus
C) Sickle cell anaemia
D) Endocarditis
E) Polyarteritis nodosa
F) IgA nephropahy

G) Systemic lupus erythematosus
H) Minimal change glomerulonephropathy
I) Membranous glmerulonephritis
J) Gold
K) Focal segmental glomerulosclerosis
L) Malaria
M) Membranoproliferative
    glomerulonephropathy

Choose the most likely cause of nephrotic syndrome from the list above that best matches each of the clinical scenarios below:

106) A 30 year old female patient has a butterfly rash. She develops acute renal failure. Biopsy reveals full house immunodeposition in the glomeruli.

107) A patient diagnosed with multiple myeloma has Bence-Jones protein in her blood and urine. She develops nephrotic syndrome.

108) A ten year old boy develops nephrotic syndrome. The renal biopsy shows normal appearances under the light microscope.

109) A female patient has glycosuria, retinopathy, renal artery stenosis and angina. She goes on to develop nephrotic syndrome. The renal biopsy reveals Kimmelstiel-Wilson lesions.

110) A businessman returns from a trip abroad with a fever and nephrotic syndrome. A blood film reveals merozoites.

A) CIN 1   F) Leiomyosarcoma   K) Atypical hyperplasia
B) Simple hyperplasia   G) CIN 3   L) Leiomyoma
C) Adenocarcinoma   H) Squamous cell carcinoma M) Endometritis
D) Brenner tumour   I) Complex hyperplasia
E) Endometriosis   J) Endometrial stromal sarcoma

Match the biopsy results below with the most likely diagnosis above.

111) An endometrial biopsy reveals invasive spindle cells that are SMA negative, CD10 positive, oestrogen receptor positive, CD146 negative and oxytocin negative.

112) An endometrial biopsy shows proliferative endometrium with foci of gland-stromal ratio of greater than 3:1, complex budding and papillary folding. No cytological atypia was seen. No invasion was present.

113) A prolapsed uterus contains a firm cream homogeneous nodule 2cm in diameter that is well circumscribed. Histology showed that the nodule contained syncytial spindle cells with cigar shaped nuclei and bland chromatin. No mitotic figures or necrosis were noted. The cells of the nodule are SMA positive.

114) An ovarian biopsy from a 50 year old woman contains invasive foci of transitional epithelium with numerous regular and irregular mitotic figures. The epithelium is CK8, CK18, EMA and CEA positive.

115) An endometrial biopsy taken on day 9 of the menstrual cycle shows proliferative endometrium and numerous diffuse neutrophils. No plasma cells are noted. A new intrauterine device has recently been installed. No hyperplasia or neoplasia are seen.

A) Squamous cell carcinoma
B) Basal cell carcinoma
C) Chondrodermatitis nodularis helicis
D) Bullous pemphigoid
E) Dermatitis herpetiformis
F) Pemphigus vulgaris

G) Pilar cyst
H) Sebaceous cyst
I) Psoriasis
J) Pilomatrixoma
K) Actinic keratosis
L) Bowen's disease
M) Epidermolysis bullosa acquisita

Match the descriptions below with the most likely dermatopathological diagnoses above.

116) Macroscopically this is a well demarcated spherical white nodule, 1.5cm in maximum extent. Microscopically the nodule is covered by a thin stratified squamous epithelium without evidence of a granular layer. The nodule contains acellular homogeneous glassy material.

117) This well defined cystic structure was found in the skin and contained basaloid cells and shadow (ghost) cells.

118) This skin lesion is characterized by basal epidermal dysplasia, parakeratosis, acanthosis and elastosis.

119) This skin lesion is classically found at the ear and is characterized by fibrinoid necrosis of cartilage or collagen.

120) This skin lesion shows mounds of parakeratosis with focal epidermal collections of neutrophils, confluent parakeratosis, hypogranulosis, spongiosis and regular epidermal hyperplasia, papillary dermal oedema and prominent dermal vessels. There is no evidence of dysplasia or malignancy.

A) Sebaceous hyperplasia
B) Sebaceous adenoma
C) Sebaceous carcinoma
D) Sebaceous epithelioma
E) Dysplastic naevus
F) Malignant melanoma

G) Psoriasis
H) Muir-Torre syndrome
I) Darier's disease
J) Focal acantholytic dyskeratosis
K) Grover's disease
L) Warty dyskeratoma

Match the descriptions below with the most likely dermatopathological diagnoses above.

121) This individual suffers from multiple sebaceous neoplasms and GI carcinomas. The disease is autosomal dominant and is linked to mutations in mismatch repair genes.

122) A 70 year old man presents with erythematous papules on his chest. Histological examination reveals suprabasal clefting with sparse dyskeratotic and acantholytic cells. There is no evidence of malignancy.

123) This isolated skin lesion is examined histologically and reveals a sebaceous appearance that is multilobulated with the sebaceous glands of various stages of maturity. Some show cystic degeneration. At least half of each lobule is composed of mature sebaceous cells. The lesion appears to connect with the surface epithelium.

124) This darkly pigmented skin lesion is 5mm in maximum extent. Histological examination reveals a lentiginous proliferation of melanocytes showing random cytological atypia. Elongation of rete ridges and adjacent ridge fusion are noted. Lamellar fibroplasia and a perivascular lymphocytic infiltrate are present.

125) A 50 year old woman presents with an isolated lesion on the skin of the head that is approximately 5mm in maximum extent. It seems to involve a hair follicle. Histological examination shows a reaction pattern that is characterized by suprabasal clefting and occasional dyskeratotic and acantholytic cells.

ANSWERS

1) C	43) D	85) C
2) D	44) E	86) H
3) C	45) A	87) O
4) C	46) B	88) A
5) B	47) B	89) B
6) B	48) E	90) M
7) B	49) A	91) I
8) D	50) B	92) P
9) B	51) F	93) Q
10) A	52) I	94) V
11) D	53) A	95) U
12) B	54) K	96) I
13) C	55) J	97) E
14) A	56) J	98) J
15) E	57) C	99) K
16) D	58) F	100) C
17) D	59) H	101) I
18) C	60) K	102) C
19) C	61) I	103) H
20) C	62) D	104) A
21) D	63) K	105) D
22) B	64) A	106) G
23) A	65) A	107) A
24) D	66) L	108) H
25) D	67) E	109) B
26) B	68) C	110) L
27) D	69) G	111) J
28) E	70) I	112) I
29) D	71) D	113) L
30) A	72) C	114) D
31) A	73) D	115) M
32) E	74) A	116) G
33) B	75) B	117) J
34) D	76) G	118) K
35) C	77) I	119) C
36) C	78) K	120) I
37) C	79) B	121) H
38) B	80) J	122) K
39) A	81) B	123) B
40) C	82) M	124) E
41) E	83) D	125) L
42) E	84) A	

Paper 5 – Answers

1) C

Britain has the lowest international rates for gallbladder cancer. Native American populations show a genetic predisposition to this disease, which is why rates are higher in South America and Alaska. There is also a high incidence among women in northern India. This appears to be due to a very high incidence of gallstones in this region.

2) D

Graft versus host disease is a **type IV** (cell mediated) immune reaction i.e. an immune response that does not involve antibodies or complement but involves the activation of macrophages, natural killer cells and antigen-specific cytotoxic T-lymphocytes.

3) C

Type III hypersensitivity reactions are characterized by soluble antigens that are not bound to cell surfaces (which is the case in type II hypersensitivity). When these antigens bind antibodies, immune complexes of different sizes form. Large complexes can be cleared by macrophages but macrophages have difficulty binding to small immune complexes for clearance. These immune complexes are deposited in heart valves, small joints, blood vessels, and glomeruli, causing symptoms.

4) C

In fact hypercalcaemia, not **hypocalcaemia**, is a cause of pancreatitis. Of scorpions, only a species found in Trinidad causes pancreatitis and there envenomation is said to be the commonest cause of pancreatitis.

5) B

A nodule size of **3mm** is generally regarded as the cut-off between micro- and macronodular cirrhosis. The latter is most often associated with chronic viral hepatitis or advanced cirrhosis of any cause.

6) B

Hepatitis B is a hepatotropic DNA virus; hence hepadnavirus. All the other hepatitis viruses above, as well as the one not mentioned, GB, are RNA viruses.

7) B

Wilson's disease is an autosomal-recessive disorder in which the liver is unable to dispose of excess dietary copper via biliary secretion. Copper accumulates, damaging liver, joints, brain (especially basal ganglia), the proximal renal tubule (causing salt wasting) and red cells (mild ongoing haemolysis is common), and making the distinctive Keyser-Fleischer corneal ring.

8) D

Pancreatic carcinoma is the odd one out in this group; it spreads via lymphatics in the usual way for carcinomas. As well as the above, yolk sac tumour also has a predilection for invading veins.

9) B

Phosphorus causes periportal necrosis, as does iron toxicity and preeclampsia. Yellow fever causes mid-zonal necrosis whilst all the others cause centrilobular necrosis. This is because there is a lower oxygen tension normally in the centrilobular zones. Agents like paracetamol and CCl_4 generate free radicals, which cause most damage where oxygen levels are low.

10) **A**

Helminths, like most parasites, characteristically cause an **eosinophilia**.

11) **D**

A keloid is formed by **scarring** (often referred to as a "keloid scar"). Depending on its maturity, a keloid is composed of mainly either type III (early) or type I (late) collagen. It is a result of an overgrowth of granulation tissue (containing collagen type III) at the site of a healed skin injury that is then slowly replaced by collagen type I. Keloids do not regress and usually recur after excision.

12) **B**

Almost by definition, Burkitt's lymphoma is associated with a chromosomal translocation of the **c-myc** gene. This gene is found at 8q24.

13) **C**

Histological examinations of positive cases show inflammation of the arterial wall with fragmentation and disruption of the **internal elastic lamina**. Multinucleated giant cells are often present, but are found in less than half of cases and are not specific for the disease.

14) **A**

The granulomas of **tuberculosis** are typically caseous ("cheesy"), in other words cell poor within the central area of necrosis, which is palely eosinophilic and amorphous. The presence of neutrophils within granulomas is suggestive of one of the other four infections listed, depending on the site and clinical picture. Cat scratch disease also gives these appearances and needs to be considered in the differential diagnosis.

15) **E**

Terminal deoxyribosyl transferase (**TdT**) is positive in over 95% of cases of acute lymphoblastic lymphoma, of both B and T-cell phenotypes. CD10 and CD34 are also frequently positive, as are CD79a or CD3 depending on the cell lineage. CD45 is typically negative. 80% of cases of lymphoblastic lymphoma are of T-cell lineage and 20% B-cell. If more than 25%, but less than 50%, blasts are present in the bone marrow then this is diagnostic of leukaemia rather than lymphoma.

16) **D**

In the rare disorder hereditary *angioedema*, bradykinin formation is caused by continuous activation of the complement system due to a deficiency in **C1-esterase inhibitor**. During attacks, the skin of the face, normally around the mouth, and the mucosa of the mouth and/or throat, as well as the tongue, swell up over a period of minutes to several hours.

17) **D**

Paraesthesias, weakness and unsteady gait may be seen in pernicious anaemia. These symptoms are due to myelin degeneration and loss of nerve fibres in the lateral and dorsal columns of the spinal cord (**posterior columns of spinal cord**).

18) **C**

Vascular invasion is a classical pattern of spread for **follicular carcinoma** of the thyroid, which often goes to the lungs and the bones. Papillary carcinoma and the other cancers listed all tend to spread to local lymph node groups.

19) **C**

Patients with **essential thrombocythaemia** frequently suffer from itching due to elevated histamine levels, which are in turn associated with basophilia. This classically occurs with bathing or showering.

20) **C**

For unknown reasons, consumption of alcohol causes chest pain in a small proportion of patients with **Hodgkin's lymphoma**, and may even be diagnostic of the condition.

21) **D**

30-45% of patients with **thymomas** have myasthenia gravis, but other autoimmune disorders may also occur; these include red cell aplasia, Addison's disease, Cushing's disease and alopecia areata.

22) **B**

CNS ischaemia sufficient to kill both neurons and glial cells, results in tissue liquefaction. **Liquefactive necrosis** may also be seen in gas gangrene in particular. Caseous necrosis is seen in TB. Coagulative necrosis is the commonest type of necrosis overall and especially in ischaemia - the cytoplasm becomes hypereosinophilic, and nuclear changes i.e. condensation (pyknosis), fragmentation (karyorrhexis), and disappearance (karyolysis) can occur. The cells persist as ghost outlines. Fibrinoid necrosis is often seen in sarcoidal granulomas and rheumatoid nodules. Enzymatic fat necrosis is seen in the setting of acute pancreatitis where lipase and other enzymes are released.

23) **A**

Although the exact benefits of a high roughage diet are disputed, there is good evidence that it is protective against diverticulosis and probably also colorectal carcinoma. It has been suggested by Denis Burkitt (*of Burkitt's lymphoma fame*) that this diet may also reduce the incidence of varicosities of the legs due to the reduced weight and hence pressure exerted. A high fibre diet does not protect against **sigmoid volvulus**.

24) **D**

Gilbert's syndrome results from a mutation of the glucuronyl transferase that conjugates and solubilizes bilirubin. It is a cause of "hepatocellular jaundice" but not cholestatic jaundice. Gilbert's syndrome may be exacerbated by other illnesses. All the others listed are causes of cholestatic jaundice.

25) **D**

Clotting factor VIII is not synthesised in the liver; all the other factors are synthesized in the liver.

26) B

Alzheimer type II glia, with swollen and pale nuclei, are characteristic of hepatic encephalopathy. The pathogenesis is multifactorial but the production of false neurotransmitters, including octopamine, by the altered gut flora seen in hepatic failure, is implicated. Spatial disorientation occurs early, as well as asterixis, and is followed by progressive mental obtundation.

27) D

All of the above except **Fabry's syndrome** are risk factors for pancreatic carcinoma. FAP is also a significant familial risk factor.

28) E

Membranoproliferative glomerulonephritis – is a type of glomerulonephritis caused by deposits in the kidney glomerular mesangium with basement membrane thickening, complement activation and glomerular damage. Most cases are caused by immune complex deposition and diabetes is not a cause.

29) D

Eosinophilic histiocytes are a classical feature of mantle cell lymphoma. Mantle cell lymphoma often shows a higher cellular turnover than other small to medium sized cell lymphomas. This is reflected in increased numbers of histiocytes phagocytosing cellular debris. All the other features are associated with marginal zone lymphomas.

30) A

Peyronie's disease is a localised fibromatosis analogous to Dupuytren's contracture of the fingers. Hypospadias and epispadias are abnormal urethral openings to the ventral and dorsal surfaces of the penis respectively. Phimosis and paraphimosis refer to the prepuce not being retractable over the penile corona and the foreskin not being replaceable to its original position post retraction, respectively.

31) A

Most cases of Sister Mary Joseph's nodule represent spread to the **umbilicus** from gastrointestinal primary tumours, especially the stomach, pancreas and colon. However approximately 25% are from primary gynaecological cancers (ovarian and uterine).

32) E

In general, lumbar puncture is not helpful in diagnosing **CNS tumours** identified on imaging, unless they actually involve the ventricular system. Lumbar puncture is indicated in all the other conditions.

33) B

Leishmaniasis is a caused by a protozoan (genus Leishmania) and is transmitted by the bite of several species of **sand fly** (subfamily Phlebotominae).

34) D

Chagas' disease is spread by the bite of the **Reduviid** or kissing bug. These emerge at night, when the inhabitants are sleeping, and transmit trypomastigotes of T. cruzi in faeces left around the bite wound.

35) C
Left ventricular hypertrophy is not a feature of tetralogy of Fallot; it can occur in systemic hypertension, heart failure and aortic stenosis.

36) C
The **Campbell de Morgan Spot** is not a painful skin lesion. The mnemonic for painful skin tumours is "EGGS LAND" i.e. eccrine spiradenoma, glomus tumour, granular cell tumour, schwannoma, leiomyoma, angiolipoma, neurofibroma and dermatofibroma.

37) C
Peutz-Jegher polyps are hamartomas hence the presence of smooth muscle.

38) B
The presence of **stromal hyalinization** does not affect prognosis. In a large clinicopathological study of gastrointestinal stromal tumours, it was determined that features associated with an adverse outcome included:

Tumour size ≥7 cm.
High cellularity.
Mucosal invasion.
High nuclear grade.
Mitotic counts ≥5/50 high power fields.
Mixed cell type.
Presence of a myxoid background and/or absence of stromal hyalinization.

39) A
Acoustic neuroma is in fact a schwannoma; it is not histologically **identical to a neurofibroma**. The trigeminal nerves may sometimes be affected, but it is very rare for a schwannoma to affect any of the other cranial nerves. Large tumours may cause increased intracranial pressure due to obstruction of the CSF pathway.

40) C
250 points must be achieved in any or all of the following categories: clinical, academic and professional, over a rolling five-year period.

41) E
National Institute for Clinical Excellence (NICE) is not involved in CPA. The fifth stakeholder group is the Association of Clinical Biochemists.

42) E
Most myelolipomas occur in the **adrenal** gland, however their great rarity means that they represent only 3% of adrenal tumours. Recent experimental evidence suggests that both the myeloid and lipomatous elements have a monoclonal origin, which strongly supports the hypothesis that myelolipomas are (benign) neoplastic lesions.

43) **D**

Consent must be explicitly obtained before **storing tissue** for any purposes, under the human tissue act of 2004.

44) **E**

CD25 expression may be seen in hairy cell leukaemia and adult T-cell lymphoma/leukaemia, but not in melanoma. CD117 and CD99 are each expressed in around 60% of cases. Over 90% express S100 and about 75% HMB45.

45) **A**

Neuroectodermal differentiation or melanocytic colonization are two alternative theories of histogenesis for the Bednar tumour, otherwise known as **pigmented dermatofibrosarcoma protuberans (DFSP)**. They show no great clinicopathological differences from conventional dermatofibrosarcoma protuberans apart from for the presence of melanocytes, representing approximately 5% of all cases of DFSP.

46) **B**

This describes a **hibernoma**, a rare benign tumor of brown fat, occurring chiefly on the back in the region of the scapulae, in adults.

47) **B**

Normally the deceased must have **been seen by a medical practitioner** within two weeks, not three months, prior to death in order for an inquest not to be opened. Other indications include any death where the cause is unknown or there are suspicious circumstances surrounding death.

48) **E**

Calretinin has been shown to be the most sensitive and specific marker for adrenal cortical carcinoma, being positive in around 96% of cases. The vast majority of cases also express the other antigens listed in the question. However **CD117** is only positive in 5% of cases.

49) **A**

A symplastic leiomyoma refers to one showing significant nuclear atypia but lacking increased mitotic figures ("**marked pleomorphism but rare mitotic figures**"). The scarcity of the latter is the key feature excluding leiomyosarcoma.

50) **B**

Osteosarcoma most commonly affects the regions around the **knee** (60% of cases). 15% of cases occur around the hip, 10% occur at the shoulder and 8% occur in the jaw.

51) **F**

Apoptosis is the process of programmed cell death. In normal growth and development, ordered growth must be balanced by programmed cell death. Hence apoptosis is ubiquitous but is particularly well exemplified by:
Endometrial breakdown during menstruation.
Regression of the enlarged breast as lactation ceases.
Hepatocyte death as a result of viral hepatitis.

Apoptosis can occur under both physiological and pathological circumstances.

52) I

Disuse **atrophy** occurs because the leg, including the ankle, is immobilized in the cast.

53) A

The disordered growth that occurs in **dysplasia** can cause of loss of nuclear polarity in the cell and disturb cellular architecture. These effects also apply to the disturbed arcitecture of whole of tissues and organs.

54) K

Either thrombolytics or angioplasty may be attempted to open an obstructed lumen in order to allow **reperfusion** of the downstream tissue. In myocardial infarction the obstruction is caused by a thrombus that is usually formed on the background atheroma.

55) J

Aplasia means the complete failure of growth. It is frequently applied to the congenital absence of an organ. A phrase such as *aplasia of the right kidney* could be used under appropriate circumstances.

56) J

This is a basic description of the first part of **meiosis**. The first division in meiosis is sometimes also called a *reduction division*; an event whereby the DNA content per cell (ploidy) is *reduced* as the cell *divides*. Meiosis is necessary to produce the germ cells – the eggs and the sperm.

57) C

DNA *replication* occurs during the **S** phase of the cell cycle.

58) F

Permanent (or non-dividing) cells and tissues cannot undergo mitotic division in adult life. Mature nerve, skeletal and cardiac muscle are such examples. Although tissue stem cells in these organs are capable of division, they are not fully differentiated.

59) H

The cell cycle is controlled by restriction points such as the G_1 to S transition. Cyclin dependent kinases become activated on binding the cofactor protein **cyclin**; the respective kinase can then phosphorylate retinoblastoma protein which causes release of the transcription factor E2F. E2F can then stimulate the transcription and synthesis of proteins important in driving the cell cycle.

60) K

G_0 is the part of the cell cycle during which quiescent cells rest and do not divide.

61) I

The presence of a single large and obstructing mass in the rectum and associated palpable lymph nodes implies a malignancy. Adeno**carcinoma** is the commonest at this site. Smoking is a risk factor for this tumour.

62) D

The outpouchings represent diverticula; a common disease in the Western world because of the low fibre diet. The presence of multiple diverticula is sufficient to describe diverticulosis. However this diverticular disease is associated with inflammation to the point of focal perforation and at least localised peritonitis. The most accurate diagnosis is **diverticulitis**.

63) K

The question stem describes the typical macroscopic appearance of the colon in a patient with **Familial Adenomatous Polyposis** (FAP). Clinically, 100 polyps are needed to make the diagnosis of FAP. The fact that the patient is only 12 years old should also cause the reader to consider FAP; this hereditary condition presnts much earlier than sporadic colorectal cancers.

64) A

The repeated bouts of bloody diarrhoea suggest that the patient has chronic idiopathic inflammatory bowel disease. The observation that the foci of disease are discontinuous, together with the presence of granulomata, suggest **Crohn's** disease.

65) A

Fat wrapping and cobblestoning are characteristic of the macroscopic appearance of **Crohn's** disease. Fat wrapping has been defined as fat hypertrophy extending from the mesenteric attachment with greater than 50% coverage of the intestinal surface. It is correlated with transmural inflammation. The cobblestone appearance occurs because of intermittent areas of inflammation separated by near-normal mucosa.

66) L

Pompe's disease occurs in individuals who have inherited a deficiency in α-1,4-**glucosidase**. The lack of this lysosomal enzyme means that glycogen cannot be broken down and accumulates in the two major sites of physiological storage; liver and muscle. This leads to heptomegaly and cardiomegaly, with disordered muscle metabolism and energetics leading to a generalized loss of muscle tone.

67) E

Ankylosing spondylitis has the strongest HLA association with **HLA B27**. The relative risk is approximately 90 (an individual with HLA B27 is 90 times more likely to have ankylosing spondylitis than an individual without HLA B27).

68) C

DiGeorge syndrome (thymic hypoplasia) is a T cell deficiency that arises out of the failure of development of the third and fourth pharyngeal pouches, leading to an under-development or lack of a thymus. Such individuals usually have a deletion on chromome **22q11**.

69) G

Rheumatoid arthritis has the strongest association with **HLA DR4**. Individuals are approximately four times more likely to have rheumatoid arthritis if they expresss HLA DR4.

70) I
BRCA-1 and BRCA-2 are both tumour suppressor genes involved in DNA repair that predispose to breast cancer if BRCA-1 or **BRCA-2** should become significantly defective. Damage to these tumour suppressor genes also predisposes to ovarian carcinomas.

71) D
The commonest type of collagen in basement membranes is **type IV**.

72) C
Ehlers-Danlos syndrome occurs because of a defect in **type III** collagen.

73) D
Goodpasture's syndrome is caused by an autoimmune attack on the collagen of the glomerular basement membrane. The anti-GBM antibodies target a component of **type IV** collagen.

74) A
Brittle bone disease is osteogenesis imperfecta. The defective collagen is **type I**; a glycine substitution impairs normal folding of this protein.

75) B
Type II collagen is present in cartilage. It is also found in vitreous humour and intervertebral disks.

76) G
The immunohistochemical panel indicated in the question stem is selective for pancreatobiliary tissue. Only **pancreas** is present here as an answer option.

77) I
Classically, **renal cell adenocarcinoma** is RCC positive, alpha beta crystallin positive, CK20 positive, CD10 positive and CK7 negative.

78) K
TTF1 is a highly specific marker for primary lung adenocarcinomas. Primary **lung** tumours are usually CK7 positive and CK20 negative.

79) B
The question stem describes a spindle cell malignancy that is positive for skeletal muscle markers; the implication is that the tumour is a rhabdomyosarcoma. MyoD1 and myoglobin are specific for **skeletal muscle**, however desmin stains both leiomyosarcoma and rhabdomyosarcoma.

80) J
PSA (prostate specific antigen) and PSAP (prostate specific acid phosphatase) are both very specific for **prostatic** adenocarcinoma primaries.

81) B

The two accepted cardiac valve causes of sudden death in adults are aortic stenosis and mitral valve prolapse. Aortic stenosis typically produces an ejection systolic murmur whereas **mitral regurgitation** causes a pansystolic murmur.

82) M

Intravenous drug users allow venous access to bacteria that most often cause infective endocarditis of the **tricuspid valve**.

83) D

Aortic regurgitation is associated with a dramatic drop in diastolic pressure as the aortic valve collapses. De Musset's sign is the visible head nodding that occurs with the drop in diastolic blood pressure. Corrigan's sign represents the prominent visible carotid pulsations that occur with wide pulse pressure. Quincke's sign represents visible capillary pulsations in the nail beds due to the wide pulse pressure. The waterhammer pulse, pistol shot femoral pulse and collapsing pulse are all signs of aortic regurgitation.

84) A

Congenital parachute valve occurs as the chordae tendinae insert into the same focus of cardiac muscle. The valve is then held in a partially closed position, causing **mitral stenosis**.

85) C

Aortic stenosis is a recognized cause of sudden death in adults and occurs due to obstruction of aortic outflow. Aortic stenosis increases in incidence with age as a result of wear and tear of the aortic valve and associated calcification. Congenital bicuspid aortic valve is predisposed to wear and tear damage and aortic stenosis.

86) H

Barrett's oesophagus is a risk factor for oesophageal adenocarcinoma; the metaplasia is the first step of a multistep oncogenesis.

87) O

Repeated miscarriages can occur in sufferers of antiphospholipid syndrome. The disease may be primary or can occur in SLE sufferers that carry **antiphospholipid antibodies.**

88) A

Transitional cell carcinoma of the bladder had a higher incidence in **aniline dye workers**. It is now believed that the carcinogenic compounds are actually naphthylamines rather than anilines.

89) B

The biggest risk factor for peptic ulceration is **Helicobacter pylori** infection. None of the other factors, A-O, are significant risk factors for peptic ulceration.

90) M

Asbestos is a risk factor for all **lung carcinomas** as well as mesotheliomas.

91) I

Thyrotoxicosis occurs when a patient is symptomatic due to excessive thyroid hormones. The commonest cause is hyperthyroidism. Hence clinically symptomatic hyperthyroidism is a description of thyrotoxicosis.

92) P

The commonest cause of Graves' disease is formation of autoantibodies to the **TSH receptor** that act as agonists at the receptor to stimulate it.

93) Q

Diffuse toxic hyperplasia causes 85% of the cases of thyrotoxicosis. Diffuse toxic hyperplasia is synonymous with Graves' disease.

94) V

Graves' disease characteristically has a clinical triad that consists of hyperthyroidism, infiltrative ophthalmopathy and infiltrative dermopathy. **Infiltrative dermopathy** is synonymous with pretibial myxoedema.

95) U

In **exophthalmos** that can occur with Graves' disease, the volume of retro-orbital connective tissues and extra-orbital muscles is increased because of all of the following:

1) Infiltration by inflammatory cells (T lymphocytes).
2) Inflammatory oedema and swelling of extraocular muscles.
3) Accumulation of hyaluronic acid and chondroitin sulphate.
4) Fatty infiltration.

96) I

Ankylosing spondylitis is a seronegative arthropathy that causes progressive kyphosis and lumbar lordosis. The concomitant calcification of the interspinous ligaments causes a characteristic X-ray appearance called bamboo spine. Ankylosing spondylitis is associated with aortic regurgitation and mitral regurgitation.

97) E

The pathogenesis of **gout** involves the deposition of sodium urate crystals in joints. These crystals can rotate the plane of polarized light and are negatively birefringent. Gout can cause both acute and chronic joint pain.

98) J

Rheumatoid arthritis is associated with HLA-DRB1 and individuals with these histocompatibility antigens are more likely to be triggered by the arthritogen that is believed to cause rheumatoid arthritis. Boutonniere and swan neck deformities are flexion-hyperextension abnormalities characteristic of rheumatoid arthritis. Rheumatoid arthritis preferentially affects the proximal interphalangeal joints with associated joint pain (arthralgia).

99) K

The low bone density that occurs with **osteoporosis** predisposes to vertebral spine fracures and distal forearm fractures. It occurs commonly in postmenopausal women.

100) C

Septic arthritis is commonest in the young and is high on the differential for acute joint pain. The preponderance of neutrophils in the aspirate is diagnostic of septic arthritis.

101) I

Churg-Strauss syndrome is a small vessel vasculitis that can cause acute severe asthma. Lung biopsy usually indicates an eosinophilia.

102) C

Kawasaki's disease is a medium vessel vasculitis that can affect the coronary arteries and often occurs in children under 5. It predisposes to coronary artery aneurysms whose rupture can cause sudden death. The disease can also cause a myocarditis or a pericarditis.

103) H

Wegener's granulomatosis is a small vessel vasculitis that has manifestations in three areas; the upper respiratory tract, the lower respiratory tract and the kidneys (if ony two of the three areas are affected then the disease is termed Limited Wegener's Granulomatosis). Sinusitis is a frequent upper respiratory tract symptom. cANCA positivity during exacerbations of the disease are virtually diagnostic for Wegener's granulomatosis.

104) A

Giant cell arteritis is a large vessel vasculitis that can affect the ophthalmic artery and is commonest in individuas over 60. It can cause blindness and is treated as a medical emergency. Headaches and scalp tenderness are a frequent part of the clinical presentation. A significant minority of the sufferers also suffer from polymyalgia rheumatica.

105) D

Buerger's disease (Thromboangiitis obliterans) is a medium vessel vasculitis that affects the arteries of the hands and feet. The typical patient is a young male who is a heavy smoker.

106) G

Systemic lupus erythematosus is commonest in females of African extraction. A butterfly rash commonly occurs in lupus. 'Full house' immunoglobulin and complement deposition is characteristic of lupus.

107) A

Multiple myeloma can produce a monoclonal protein to cause **amyloidosis**. Bence-Jones protein is the monoclonal protein produced by this malignancy to cause AL amyloidosis. It is derived from immunoglobulin light chains.

108) H

Minimal change glomerulonephropathy is the commonest cause of nephrotic syndrome in childhood. The podocyte foot process effacement is the most characteristic change visible under microscopy – however it cannot be seen under light microscopy but can be seen under *electron microscopy*.

109) B

The glycosuria is suggestive of **diabetes mellitus**. Kimmelstiel-Wilson lesions are a feature of diabetic nephropathy and retinopathy can occur as a chronic complication of diabetes mellitus. Renal artery stenosis and angina are likely to be as a result of atherosclerosis – another consequence of diabetes mellitus.

110) L

Malaria caused by Plasmodium sp. can cause a chronic infection after being transmitted by its vector; the female Anopheles mosquito. Merozoites can be seen in the blood of infected individuals. One complication of this infection is nephrotic syndrome.

111) J

The differential for a primary spindle cell malignancy of the uterus includes leiomyosarcoma and **endometrial stromal sarcoma**. Endometrial stromal sarcoma is usually CD10 positive and positive for oestrogen (and progesterone) receptors. Classically, leiomyosarcoma is SMA positive. It may be CD146 positive and oxytocin positive also.

112) I

The question stem describes excessive hyperplasia of endometrial glands such that the gland-stromal ratio is greater than 3:1. This is associated with gland architectural atypia but not cytological atypia. This is the description of **complex hyperplasia**.

113) L

The macroscopic and microscopic descriptions are typical of a **leiomyoma**. No invasion, necrosis or mitotic figures were found so the small possibility of a leiomyosarcoma has been excluded.

114) D

The invasive transitional epithelium is likely to be a **Brenner tumour**. The immunoprofile is consistent with this diagnosis.

115) M

The endometrial biopsy is acutely inflamed and is from a woman who has an intrauterine device. The most likely diagnosis is **endometritis**.

116) G

The question stem describes a cyst-like structure whose epithelium lacks a granular layer. The cyst contains an acellular glassy material consistent with keratin. The description is of a **pilar cyst**. A sebaceous cyst has laminated keratin and a granular layer is generally present.

117) J

Basloid cells and shadow or ghost cells are found in a **pilomatrixoma**.

118) K

The question stem describes the core features of an **actinic keratosis**. If the dysplasia had been severe and involved the full thickness of the epidermis then IEC (intraepidermal carcinoma) would be the diagnosis.

119) C

Chondrodermatitis nodularis helicis is often found in males over 40. A well-formed lesion can demonstrate overlying ulceration, parakeratosis, hyperkeratosis, fibrinoid necrosis of cartilage/collagen, granulation tissue and inflammatory cells.

120) I

The question stem describes collections of neutrophils in the keratin layer, this is consistent with Munro microabscesses. The regular epidermal hyperplasia is consistent with regular psoriasiform hyperplasia. The most appropriate diagnosis in the offered selection is **psoriasis**.

121) H

The **Muir-Torre syndrome** predisposes to multiple sebaceous neoplasms and GI carcinomas. The underlying defects are mutations in the mismatch repair genes hMLH1 and hMSH2 together with microsatellite instability.

122) K

The question stem describes acantholytic dyskeratosis as a feature of this dermatopathology. The different types require some knowledge of the clinical presentation to allow discrimination. **Grover's disease** involves multiple lesions that tend to present on the trunks of older men.

123) B

The key features in the description are the sebaceous character, the glands of various stages of maturity and the solitary nature of this skin lesion. These are the features of a **sebaceous adenoma**.

124) E

This proliferation of atypical melanocytes does not show epidermal or dermal invasion. It is therefore more likely to be a **dysplastic naevus** than a malignant melanoma.

125) L

The question stem describes another acantholytic dyskeratosis. This is an isolated lesion on the head of a middle-aged individual that involves a hair follicle. This clinical history is typical of **warty dyskeratoma**.

Lightning Source UK Ltd.
Milton Keynes UK
UKOW07f2154310715

256204UK00011B/183/P